SpringerBriefs in Mathematical Physics

Volume 1

Series editors

Nathanaël Berestycki, Cambridge, UK
Mihalis Dafermos, Cambridge, UK
Tohru Eguchi, Tokyo, Japan
Atsuo Kuniba, Tokyo, Japan
Matilde Marcolli, Pasadena, USA
Bruno Nachtergaele, Davis, USA

For further volumes:
http://www.springer.com/series/11953

Yoshiaki Tanii

Introduction to Supergravity

 Springer

Yoshiaki Tanii
Graduate School of Science
   and Engineering
Saitama University
Saitama
Japan

ISSN 2197-1757   ISSN 2197-1765 (electronic)
ISBN 978-4-431-54827-0   ISBN 978-4-431-54828-7 (eBook)
DOI 10.1007/978-4-431-54828-7
Springer Tokyo Heidelberg New York Dordrecht London

Library of Congress Control Number: 2014940707

Printed on acid-free paper

Springer is part of Springer Science+Business Media (www.springer.com)

Preface

This book is a pedagogical introduction to supergravity. Supergravity is a gravitational field theory that includes supersymmetry (symmetry between bosons and fermions) and is a generalization of Einstein's general relativity. Supergravity provides a low energy effective theory of superstring theory, which has attracted much attention as a candidate for the unified theory of fundamental particles, and it is a useful tool for studying nonperturbative properties of superstring theory such as D-branes and string duality.

This work considers classical supergravities in four and higher dimensional spacetime with their applications to superstring theory in mind. More concretely, it discusses classical Lagrangians (or field equations) and symmetry properties of supergravities. Besides local symmetries, supergravities often have global non-compact symmetries, which play a crucial role in their applications to superstring theory. One of the main features of this book is its detailed discussions of these non-compact symmetries.

The aim of the book is twofold. One is to explain the basic ideas of supergravity to those who are not familiar with it. Toward that end, the discussions are made both pedagogical and concrete by stating equations explicitly. The other is to collect relevant formulae in one place so as to be useful for applications to string theory. They include the lists of possible types of spinors in each dimension, field contents of supergravities, and global symmetries of supergravities.

Most of the discussions are restricted to pure supergravities without matter couplings. An exception is a coupling to the super Yang–Mills multiplet in ten dimensions. Supergravities in lower than four dimensions are not considered. There are many other issues on supergravity which are not discussed in this book. For those issues, consult the references given at the end of Sect. 1.1.

The plan of the book is as follows. In Chap. 1, we first explain a role of supergravity in superstring theory briefly, and then review the formulations of the gravitational field and other fields coupled to it. In Chap. 2, we discuss supergravities in four dimensions in details. Much of the properties of supergravities in higher dimensions already appear in four dimensions. In Chap. 3, we discuss superalgebras and supermultiplets in general dimensions and give the lists of possible types of supergravities and their field contents. In Chap. 4, we consider global non-compact symmetries in supergravities, which are useful in understanding the structure of scalar fields. The non-compact symmetry in supergravities is sometimes realized as a duality symmetry of vector or antisymmetric tensor fields, which is a generalizations of the electric–magnetic duality in Maxwell's equations. In Chap. 5, we consider supergravities in higher dimensions. In particular, the Lagrangians (or field equations) and the symmetry properties of supergravities in 11 and 10 dimensions are discussed in details. In Chap. 6, we consider dimensional reductions of supergravities in eleven and ten dimensions to lower dimensions in order to understand the origins of the global non-compact symmetries. Finally, in Chap. 7, we consider gauged supergravities, which have minimal couplings to vector gauge fields, and massive supergravities similar to them. Notation and conventions used in this book are summarized in Appendix A. Formulae of gamma matrices and spinors in general dimensions are collected in Appendix B.

Saitama, April 2014   Yoshiaki Tanii

Contents

Chapter 1
Introduction

1.1 Supergravity and Superstring

Supergravity is a theory of gravity which has supersymmetry, a symmetry between bosons and fermions. Supersymmetry in supergravity is a local symmetry like the gauge symmetry in the standard theory of particle physics. The gauge field of the local supersymmetry is the Rarita–Schwinger field, which represents a particle with spin $\frac{3}{2}$ called a gravitino. Supergravity also has a local symmetry under the general coordinate transformation, whose gauge field is the gravitational field.

An important role of supergravity is in its relation to superstring theory. Superstring theory is a candidate of the unified theory of fundamental particles including gravity. It is expected that superstring theory provides a consistent quantum theory of gravity since it has no ultraviolet divergence. Superstring theory contains massless states such as graviton and gauge particles in addition to infinitely many massive states. The mass scale of the theory is supposed to be very large, say the Planck energy scale, and only the massless states are important at low energy. Supergravity provides a low energy effective theory of the massless sector of superstring theory and can be used to study its low energy properties.

Superstring theory is an incomplete theory at present as its fundamental formulation is not known. As a consequence, it is difficult to study its non-perturbative properties, whose understanding is indispensable for applications to physics. In the 1990s there was considerable progress in non-perturbative understanding of string theory, which was possible by two important discoveries: D-branes [8] and string duality [4, 16].

D-branes are spatially extended objects appearing in string theory and correspond to solitons in field theories. As solitons are useful tools in non-perturbative study of field theories, D-branes can be used to study non-perturbative properties of string theory. At low energy, D-branes are represented by classical solutions of the effective supergravity theory. One can study various aspects of D-branes by using supergravity.

String duality is the relation among seemingly different string theories. Five super-symmetric string theories are known. They are type I, type IIA, type IIB superstring

Y. Tanii, *Introduction to Supergravity*, SpringerBriefs in Mathematical Physics, DOI: 10.1007/978-4-431-54828-7_1, © The Author(s) 2014

theories and two kinds of heterotic string theories, all of which are formulated in 10-dimensional spacetime. String duality provides relations among these superstring theories. Furthermore, it predicts another theory in 11-dimensional spacetime called M theory [11, 16] as a strong coupling limit of type IIA superstring theory. By string duality these six theories are intimately related, and it is conjectured that they are different aspects of one fundamental theory. String duality sometimes relates a theory at strong coupling and another theory at weak coupling. One can use this relation to study the former in terms of the latter, which may be studied by a perturbative method. Thus, string duality is extremely useful to understand non-perturbative properties of string theories. However, at present understanding of string theory, it is difficult to prove string duality directly in full string theory. On the other hand, the massless sectors of superstring theories are described by supergravities, for which complete field theoretic formulations are known at the classical level. One can obtain information on string duality by using supergravity. Indeed, supergravities containing scalar fields have a global symmetry of non-compact Lie group, whose discrete subgroup corresponds to the expected group of string duality.

In this book we consider classical supergravities in four and higher dimensional spacetime with their applications to superstring theory and M theory in mind. More concretely, we discuss classical Lagrangians (or field equations) of supergravities and their symmetry properties. In particular, we will discuss global non-compact symmetries in detail, which are related to string duality as mentioned above. We do not try to give complete references to the original work. For more complete references see [2, 10]. Other useful references on supergravity are [6, 13, 14]. For reviews of superstring theory and M theory, see [1, 3, 9, 17].

1.2 Gravitational Field

Supergravity contains various kinds of fields including the gravitational field. In the following we review the formulations of the gravitational field and other fields coupled to it. We consider field theories in general D-dimensional spacetime. Spacetime coordinates are denoted as x^μ ($\mu = 0, 1, \cdots, D - 1$), where x^0 is time. It is convenient to use the variational principle to formulate field theories. The action S is an integral of the Lagrangian (density) $\mathscr{L}$ over D-dimensional spacetime

$$S = \int d^D x \, \mathscr{L}, \qquad d^D x = dx^0 dx^1 \ldots dx^{D-1}. \tag{1.1}$$

The Lagrangian $\mathscr{L}$ is a function of the fields and their derivatives. The field equations are obtained by the least action principle for S.

There are two formulations of the gravitational field: the metric formulation and the vielbein formulation. To couple gravity to tensor fields such as scalar and vector fields, both formulations can be used. To couple gravity to spinor fields, the vielbein formulation should be used.

1.2.1 Metric Formulation

In the usual formulation of the gravitational theory a metric $g_{\mu\nu}(x)$ is used to describe gravity. Such a formulation is called the metric formulation. The Lagrangian for the gravitational field $g_{\mu\nu}(x)$ is

$$\mathscr{L} = \frac{1}{16\pi G} \sqrt{-g} \, (R - 2\Lambda), \tag{1.2}$$

where G is Newton's gravitational constant, Λ is the cosmological constant and $g = \det g_{\mu\nu}$. The first and the second terms are called the Einstein term and the cosmological term, respectively. In the following we will set $16\pi G = 1$ for simplicity. The scalar curvature R is defined from the Ricci tensor $R_{\mu\nu}$ and the Riemann tensor $R_{\mu\nu}{}^{\rho}{}_{\sigma}$ as

$$R = g^{\mu\nu} R_{\mu\nu}, \qquad R_{\mu\nu} = R_{\rho\mu}{}^{\rho}{}_{\nu},$$
$$R_{\mu\nu}{}^{\rho}{}_{\sigma} = \partial_{\mu}\Gamma^{\rho}_{\nu\sigma} - \partial_{\nu}\Gamma^{\rho}_{\mu\sigma} + \Gamma^{\rho}_{\mu\lambda}\Gamma^{\lambda}_{\nu\sigma} - \Gamma^{\rho}_{\nu\lambda}\Gamma^{\lambda}_{\mu\sigma}, \tag{1.3}$$

where $g^{\mu\nu}$ is the inverse matrix of $g_{\mu\nu}$. The Christoffel symbol $\Gamma^{\lambda}_{\mu\nu}$ is defined as

$$\Gamma^{\lambda}_{\mu\nu} = \frac{1}{2} g^{\lambda\rho} \left(\partial_{\mu} g_{\nu\rho} + \partial_{\nu} g_{\mu\rho} - \partial_{\rho} g_{\mu\nu} \right). \tag{1.4}$$

This form of the Christoffel symbol is uniquely determined by the two conditions

$$\text{metricity condition: } \partial_{\lambda} g_{\mu\nu} - \Gamma^{\rho}_{\lambda\mu} g_{\rho\nu} - \Gamma^{\rho}_{\lambda\nu} g_{\mu\rho} = 0,$$
$$\text{torsionless condition: } \Gamma^{\lambda}_{\mu\nu} = \Gamma^{\lambda}_{\nu\mu}. \tag{1.5}$$

By using the Christoffel symbol we can define the covariant derivative D_{μ} on tensor fields. For instance, the covariant derivatives on a contravariant vector field V^{μ} and a covariant vector field V_{μ} are defined as

$$D_{\mu} V^{\nu} = \partial_{\mu} V^{\nu} + \Gamma^{\nu}_{\mu\rho} V^{\rho},$$
$$D_{\mu} V_{\nu} = \partial_{\mu} V_{\nu} - \Gamma^{\rho}_{\mu\nu} V_{\rho}. \tag{1.6}$$

By using the covariant derivative, the metricity condition in (1.5) can be written as $D_{\lambda} g_{\mu\nu} = 0$. The covariant derivatives of tensor fields transform covariantly under the general coordinate transformation given below.

The field equation of $g_{\mu\nu}(x)$ derived from the Lagrangian (1.2) by the variational principle is the Einstein equation with a cosmological term

$$R_{\mu\nu} - \frac{1}{2} g_{\mu\nu} R + \Lambda g_{\mu\nu} = 8\pi G \, T_{\mu\nu}. \tag{1.7}$$

$T_{\mu\nu}$ is the energy–momentum tensor of matter fields. For now, $T_{\mu\nu} = 0$ as there is no matter field.

The action (1.1) for the Lagrangian (1.2) is invariant under the general coordinate transformation. The general coordinate transformation of the gravitational field $g_{\mu\nu}(x) \to g'_{\mu\nu}(x)$ corresponding to a change of the coordinates $x^{\mu} \to x'^{\mu} = x'^{\mu}(x)$ is determined by

$$g'_{\mu\nu}(x') = \frac{\partial x^{\rho}}{\partial x'^{\mu}} \frac{\partial x^{\sigma}}{\partial x'^{\nu}} g_{\rho\sigma}(x). \tag{1.8}$$

For $x'^{\mu} = x^{\mu} - \xi^{\mu}(x)$, where $\xi^{\mu}(x)$ is an arbitrary infinitesimal vector function, the variation of the gravitational field $\delta_G g_{\mu\nu}(x) = g'_{\mu\nu}(x) - g_{\mu\nu}(x)$ is given by

$$\begin{aligned} \delta_G g_{\mu\nu} &= \xi^{\rho} \partial_{\rho} g_{\mu\nu} + \partial_{\mu} \xi^{\rho} g_{\rho\nu} + \partial_{\nu} \xi^{\rho} g_{\mu\rho} \\ &= D_{\mu} \xi_{\nu} + D_{\nu} \xi_{\mu}. \end{aligned} \tag{1.9}$$

The variation of the Lagrangian (1.2) under this transformation is a total divergence

$$\delta_G \mathscr{L} = \partial_{\mu} \left(\xi^{\mu} \mathscr{L} \right) \tag{1.10}$$

and therefore the action (1.1) is invariant when an appropriate boundary condition is imposed on the field at infinity.

Coupling to a Scalar Field

As a matter field, let us first consider a real scalar field $\phi(x)$. The Lagrangian of a free scalar field of mass m coupled to the gravitational field is

$$\mathscr{L} = \sqrt{-g} \left[-\frac{1}{2} g^{\mu\nu} \partial_{\mu}\phi \partial_{\nu}\phi - \frac{1}{2} m^2 \phi^2 \right]. \tag{1.11}$$

Under the general coordinate transformation (1.9) and

$$\delta_G \phi = \xi^{\mu} \partial_{\mu} \phi \tag{1.12}$$

the Lagrangian transforms to a total divergence as in (1.10) and the action is invariant.

1.2.2 Vielbein Formulation

To couple gravity to spinor fields we have to use the vielbein formulation of gravity. In this formulation we introduce D independent vectors $e_a{}^{\mu}(x)$ ($a = 0, 1, \ldots, D-1$) at each point of spacetime, which are orthogonal to each other and have a unit length

$$e_a{}^{\mu}(x) e_b{}^{\nu}(x) g_{\mu\nu}(x) = \eta_{ab}, \qquad \eta_{ab} = \text{diag}(-1, +1, \ldots, +1). \tag{1.13}$$

We also introduce the inverse matrix $e_\mu{}^a(x)$, which satisfies

$$e_\mu{}^a(x)e_a{}^v(x) = \delta_\mu^v, \qquad e_a{}^\mu(x)e_\mu{}^b(x) = \delta_a^b. \tag{1.14}$$

The field $e_\mu{}^a(x)$ is called the vielbein (vierbein or tetrad in four dimensions, fünfbein in five dimensions, etc.). From (1.13) and (1.14) we can express the metric in terms of the vielbein as

$$g_{\mu v}(x) = e_\mu{}^a(x)e_v{}^b(x)\eta_{ab}. \tag{1.15}$$

Therefore, we can use the vielbein $e_\mu{}^a(x)$ as dynamical variables representing the gravitational field. We choose $e_\mu{}^a$ such that the determinant $e = \det e_\mu{}^a$ is positive and therefore $\sqrt{-g} = e$.

For a given metric $g_{\mu v}$ the vielbein $e_\mu{}^a$ satisfying (1.15) is not uniquely determined. If $e_\mu{}^a$ satisfies (1.15), then

$$e'_\mu{}^a(x) = e_\mu{}^b(x)\Lambda_b{}^a(x), \qquad \Lambda_a{}^c(x)\Lambda_b{}^d(x)\eta_{cd} = \eta_{ab} \tag{1.16}$$

also satisfies (1.15) with the same $g_{\mu v}$. The vielbein has D^2 independent components while the metric has $\frac{1}{2}D(D+1)$ independent components. The difference $\frac{1}{2}D(D-1)$ is the number of independent components of $\Lambda_a{}^b$ in (1.16). The transformation (1.16) is called the local Lorentz transformation. A theory originally given in the metric formulation should be invariant under the local Lorentz transformation when it is rewritten in the vielbein formulation. Thus, the gravitational theory in the vielbein formulation has two local symmetries: the symmetries under the general coordinate transformation and the local Lorentz transformation.

We have now two kinds of vector indices: $\mu, v, \ldots$ and $a, b, \ldots$. To distinguish them, the indices $\mu, v, \ldots$ are called the world indices, while $a, b, \ldots$ are called the local Lorentz indices. These two kinds of indices are converted into each other by using the vielbein and its inverse, e.g., $V_a(x) = e_a{}^\mu(x)V_\mu(x)$, $V_\mu(x) = e_\mu{}^a(x)V_a(x)$. Raising and lowering of indices are done by using the metric $g^{\mu v}, g_{\mu v}$ for world indices and the flat metric η^{ab}, η_{ab} for local Lorentz indices, e.g., as $V^\mu = g^{\mu v}V_v$, $V^a = \eta^{ab}V_b$. Tensor fields with local Lorentz indices transform under the local Lorentz transformation as in (1.16). They also transform under the general coordinate transformation as tensor fields according to the world indices they have. For instance, the general coordinate (G) and the local Lorentz (L) transformations of the vielbein $e_\mu{}^a(x)$ are

$$\delta_G e_\mu{}^a = \xi^v \partial_v e_\mu{}^a + \partial_\mu \xi^v e_v{}^a, \quad \delta_L e_\mu{}^a = -\lambda^a{}_b e_\mu{}^b, \tag{1.17}$$

where $\xi^\mu(x)$ and $\lambda_{ab}(x) = -\lambda_{ba}(x)$ are infinitesimal transformation parameters. The relation of $\lambda^a{}_b$ to $\Lambda^a{}_b$ in (1.16) is $\Lambda^a{}_b = \delta_b^a + \lambda^a{}_b$.

To construct the covariant derivative for the local Lorentz transformation we need a gauge field. The gauge field of the local Lorentz transformation is called the spin connection $\omega_\mu{}^a{}_b(x)$ $(\omega_{\mu ab} = -\omega_{\mu ba})$, which transforms as

$$\delta_L \omega_\mu{}^a{}_b = D_\mu \lambda^a{}_b \equiv \partial_\mu \lambda^a{}_b + \omega_\mu{}^a{}_c \lambda^c{}_b - \lambda^a{}_c \omega_\mu{}^c{}_b. \qquad (1.18)$$

The covariant derivative acting on, e.g., a vector V^a is defined as

$$D_\mu V^a = \partial_\mu V^a + \omega_\mu{}^a{}_b V^b. \qquad (1.19)$$

Under the local Lorentz transformation $D_\mu V^a$ transforms in the same way as V^a.

The Christoffel symbol $\Gamma^\lambda_{\mu\nu}$ is not an independent field but is determined by the metric as in (1.4) by imposing the metricity and torsionless conditions (1.5). Similarly, the spin connection is completely determined by the vielbein if we impose the torsionless condition

$$D_\mu e_\nu{}^a - D_\nu e_\mu{}^a \equiv \partial_\mu e_\nu{}^a + \omega_\mu{}^a{}_b e_\nu{}^b - (\mu \leftrightarrow \nu) = 0. \qquad (1.20)$$

Since the indices μ and ν are antisymmetrized, the Christoffel symbols in the covariant derivatives are canceled each other. (The metricity condition corresponds to the antisymmetry property $\omega_{\mu ab} = -\omega_{\mu ba}$, which we assumed from the start.) The unique solution of (1.20) is

$$\omega_{\mu ab}(e) = \frac{1}{2} \left(e_a{}^\nu \Omega_{\mu\nu b} - e_b{}^\nu \Omega_{\mu\nu a} - e_a{}^\rho e_b{}^\sigma e_\mu{}^c \Omega_{\rho\sigma c} \right),$$
$$\Omega_{\mu\nu a} = \partial_\mu e_{\nu a} - \partial_\nu e_{\mu a}. \qquad (1.21)$$

We can check that the local Lorentz transformation of the vielbein in (1.17) induces the transformation of $\omega_{\mu ab}$ in (1.18).

The spin connection (1.21) is related to the Christoffel symbol (1.4) as

$$D_\mu e_\nu{}^a \equiv \partial_\mu e_\nu{}^a + \omega_\mu{}^a{}_b e_\nu{}^b - \Gamma^\lambda_{\mu\nu} e_\lambda{}^a = 0. \qquad (1.22)$$

From this relation we can show that the field strength of the spin connection

$$R_{\mu\nu}{}^a{}_b = \partial_\mu \omega_\nu{}^a{}_b - \partial_\nu \omega_\mu{}^a{}_b + \omega_\mu{}^a{}_c \omega_\nu{}^c{}_b - \omega_\nu{}^a{}_c \omega_\mu{}^c{}_b \qquad (1.23)$$

is related to the Riemann tensor in (1.3) as

$$R_{\mu\nu}{}^\rho{}_\sigma = R_{\mu\nu}{}^a{}_b e_a{}^\rho e_\sigma{}^b. \qquad (1.24)$$

$R_{\mu\nu}{}^a{}_b$ is also called the Riemann tensor.

The transformation of the gravitational field

$$e_\mu{}^a(x) \rightarrow e'_\mu{}^a(x) = e^{\sigma(x)} e_\mu{}^a(x), \qquad (1.25)$$

where $\sigma(x)$ is an arbitrary function, is called the Weyl transformation. Under the Weyl transformation the spin connection and the scalar curvature transform as

$$\omega'_{\mu ab} = \omega_{\mu ab} + 2e_{\mu[a}\partial_{b]}\sigma,$$
$$R' = e^{-2\sigma}\left[R - 2(D-1)D^\mu\partial_\mu\sigma - (D-1)(D-2)\partial_\mu\sigma\partial^\mu\sigma\right]. \quad (1.26)$$

We will use these formulae in Chap. 6.

Coupling to a Spinor Field

A spinor field $\psi(x)$ transforms as a scalar field under the general coordinate transformation. Under the local Lorentz transformation it transforms as

$$\delta_L\psi = -\frac{1}{4}\lambda_{ab}\gamma^{ab}\psi, \quad (1.27)$$

where $\gamma^{ab} = \frac{1}{2}(\gamma^a\gamma^b - \gamma^b\gamma^a)$ is the antisymmetrized product of two gamma matrices. The gamma matrices γ^a in D dimensions are $2^{[D/2]} \times 2^{[D/2]}$ constant matrices satisfying the anticommutation relation

$$\{\gamma^a, \gamma^b\} = 2\eta^{ab}\,\mathbf{1}. \quad (1.28)$$

Here, $[D/2]$ denotes the largest integer not greater than $D/2$, and $\mathbf{1}$ is the unit matrix. Correspondingly, spinors in D dimensions have $[D/2]$ components. See Sect. 3.1 for more details of gamma matrices and spinors in general dimensions.

The covariant derivative of a spinor field $\psi(x)$ is

$$D_\mu\psi = \left(\partial_\mu + \frac{1}{4}\omega_\mu{}^{ab}\gamma_{ab}\right)\psi, \quad (1.29)$$

which transforms covariantly under both of the general coordinate and local Lorentz transformations. From (1.18), (1.27) the local Lorentz transformation of $D_\mu\psi$ has the same form as (1.27). The commutator of two covariant derivatives acting on a spinor field produces the Riemann tensor (1.23) as

$$[D_\mu, D_\nu]\psi = \frac{1}{4}R_{\mu\nu}{}^{ab}\gamma_{ab}\psi. \quad (1.30)$$

The Lagrangian of a free spinor field of mass m coupled to the gravitational field is

$$\mathcal{L} = -e\,\bar\psi\gamma^\mu D_\mu\psi - me\bar\psi\psi, \quad (1.31)$$

where $\bar\psi = \psi^\dagger i\gamma^0$ is the Dirac conjugate of ψ. The gamma matrices with a world index γ^μ are defined as $\gamma^\mu = \gamma^a e_a{}^\mu$. While γ^a are constant, γ^μ depend on the spacetime coordinates x^μ through $e_a{}^\mu(x)$. Note that γ^0 in the Dirac conjugate is a constant matrix $\gamma^{a=0}$. This Lagrangian is invariant under the local Lorentz transformation (1.17) and (1.27). It is also invariant under the general coordinate transformation up to a total divergence as in (1.10), and therefore the action is invariant. Spinor fields are treated as anticommuting quantities (Grassmann numbers) at the classical level since the canonical anticommutation relations are imposed on them when quantized. The complex conjugate of a product of spinor fields reverses the order of the fields as $(\psi_1 \psi_2)^* = \psi_2^* \psi_1^*$. According to this definition, the action made from the Lagrangian (1.31) is real $S^* = S$.

1.3 Yang–Mills Field

The Yang–Mills field of a gauge group G consists of vector fields $A_\mu^I(x)$ ($I = 1, 2, \cdots, \dim G$). It is convenient to express the Yang–Mills field by a matrix-valued field

$$A_\mu(x) = -ig A_\mu^I(x) T_I, \tag{1.32}$$

where g is a gauge coupling constant. T_I are representation matrices of the Lie algebra of G and satisfy the commutation relation and the normalization condition

$$[T_I, T_J] = i f_{IJ}{}^K T_K, \qquad \mathrm{tr}(T_I T_J) = \frac{1}{2} \delta_{IJ}, \tag{1.33}$$

where $f_{IJ}{}^K$ is the structure constant of the Lie algebra. In terms of the field (1.32) the infinitesimal gauge transformation is given by

$$\delta_g A_\mu = D_\mu v \equiv \partial_\mu v + [A_\mu, v], \qquad v(x) = -i v^I(x) T_I, \tag{1.34}$$

where $v^I(x)$ are arbitrary infinitesimal transformation parameters. The field strength of the Yang–Mills field is defined as

$$F_{\mu\nu} = \partial_\mu A_\nu - \partial_\nu A_\mu + [A_\mu, A_\nu] = -ig F_{\mu\nu}^I T_I. \tag{1.35}$$

Under the gauge transformation (1.34) it transforms covariantly as

$$\delta_g F_{\mu\nu} = -[v, F_{\mu\nu}]. \tag{1.36}$$

The Lagrangian of the Yang–Mills field coupled to the gravitational field is

$$\mathscr{L} = \frac{1}{2g^2} e\, g^{\mu\rho} g^{\nu\sigma} \mathrm{tr}(F_{\mu\nu} F_{\rho\sigma}) = -\frac{1}{4} e\, g^{\mu\rho} g^{\nu\sigma} F_{\mu\nu}^I F_{\rho\sigma}^I. \tag{1.37}$$

This Lagrangian is invariant under the gauge transformation (1.34). It is also invariant up to total divergences under the general coordinate transformation.

Interactions of the Yang–Mills field with other fields are introduced by using the gauge covariant derivative. Such interactions are called the minimal coupling. As an example let us consider a spinor field $\psi(x)$ which belongs to a certain representation of the gauge group. Under the gauge transformation it transforms as

$$\delta_g \psi = -v\psi, \tag{1.38}$$

where the transformation parameter v is given by (1.34) with T_I in the representation that ψ belongs to. The covariant derivative on this spinor field contains the Yang–Mills field in addition to the spin connection

$$D_\mu \psi = \left(\partial_\mu + \frac{1}{4}\omega_{\mu ab}\gamma^{ab} + A_\mu\right)\psi, \tag{1.39}$$

where T_I in (1.32) are the same as those of v in (1.38). This covariant derivative transforms in the same way as (1.38) under the gauge transformations (1.34), (1.38). Hence, the Lagrangian (1.31) is gauge invariant.

1.4 Antisymmetric Tensor Field

Supergravities in higher dimensions contain antisymmetric tensor fields of various ranks. Let us consider an antisymmetric tensor field of n-th rank $B_{\mu_1\mu_2...\mu_n}(x)$. This includes special cases of a scalar field for $n = 0$ and a vector field for $n = 1$.

The field strength of the antisymmetric tensor field is the $(n + 1)$-th rank tensor

$$F_{\mu_1\mu_2...\mu_{n+1}} = (n+1)\partial_{[\mu_1}B_{\mu_2...\mu_{n+1}]}, \tag{1.40}$$

which is invariant under the gauge transformation

$$\delta_g B_{\mu_1...\mu_n} = n\,\partial_{[\mu_1}\zeta_{\mu_2...\mu_n]}. \tag{1.41}$$

The transformation parameter $\zeta_{\mu_1...\mu_{n-1}}(x)$ is an arbitrary function totally antisymmetric in its $n - 1$ indices. The field strength satisfies the Bianchi identity

$$\partial_{[\mu_1}F_{\mu_2...\mu_{n+2}]} = 0. \tag{1.42}$$

This identity is a necessary condition for the field strength being expressed by the field $B_{\mu_1...\mu_n}$ as in (1.40). The Hodge dual of the field strength is defined by

$$(*F)^{\mu_1...\mu_{D-n-1}} = \frac{1}{(n+1)!}e^{-1}\varepsilon^{\mu_1...\mu_{D-n-1}\nu_1...\nu_{n+1}}F_{\nu_1...\nu_{n+1}}, \tag{1.43}$$

where $\varepsilon^{\mu_1\cdots\mu_D}$ is the totally antisymmetric Levi-Civita symbol with a component $\varepsilon^{012\ldots D-1} = +1$ (see Appendix A). Then, the Bianchi identity (1.42) can be rewritten as

$$\partial_{\mu_1}\left(e(*F)^{\mu_1\cdots\mu_{D-n-1}}\right) = 0. \tag{1.44}$$

The Lagrangian of a free massless antisymmetric tensor field coupled to gravity is

$$\mathcal{L} = -\frac{1}{2(n+1)!}eF_{\mu_1\ldots\mu_{n+1}}F^{\mu_1\cdots\mu_{n+1}}, \tag{1.45}$$

which is called the Maxwell type Lagrangian. The field equation derived from this Lagrangian is

$$\partial_{\mu_1}\left(eF^{\mu_1\cdots\mu_{n+1}}\right) = 0. \tag{1.46}$$

In the special cases of $n = 0$ and $n = 1$ this equation becomes the Klein–Gordon equation for a massless scalar field and Maxwell's equations for a massless vector field.

The Bianchi identity (1.44) and the field equation (1.46) have a similar form. In particular, for an antisymmetric tensor field of rank $n = \frac{1}{2}D - 1$ in even D dimensions, the field strength and its Hodge dual have the same rank $\frac{1}{2}D$, and we can consider a symmetry between them. This is a generalization of the symmetry between the electric and magnetic fields in Maxwell's equations. Such a symmetry is called the duality symmetry and plays an important role in supergravity and superstring theories. We will discuss the duality symmetry in detail in Sect. 4.2.

1.4.1 Dual Field

Because of the gauge invariance the number of physical degrees of freedom of a massless antisymmetric tensor field of rank n in D dimensions is $_{D-2}C_n$, which is the number of its transverse components $B_{i_1\ldots i_n}$ $(i_1, \ldots, i_n = 1, 2, \ldots, D-2)$. Since $_{D-2}C_n = {}_{D-2}C_{D-n-2}$, the numbers of physical degrees of freedom of antisymmetric tensor fields of rank n and rank $D - n - 2$ are the same. In certain cases a theory of an antisymmetric tensor field of rank n can be rewritten by using an antisymmetric tensor field of rank $D - n - 2$.

Let us consider an antisymmetric tensor field of rank n with the Lagrangian

$$\mathcal{L} = -\frac{1}{2(n+1)!}eF_{\mu_1\ldots\mu_{n+1}}F^{\mu_1\cdots\mu_{n+1}} + \frac{1}{(n+1)!}eF_{\mu_1\ldots\mu_{n+1}}H^{\mu_1\cdots\mu_{n+1}}, \tag{1.47}$$

where $H_{\mu_1...\mu_{n+1}}$ is an antisymmetric tensor of rank $n+1$ made of other fields. We have assumed here that the antisymmetric tensor field $B_{\mu_1...\mu_n}$ appears in the Lagrangian only through its field strength. The field equation derived from this Lagrangian and the Bianchi identity are

$$\partial_{\mu_1}\left(e(F-H)^{\mu_1...\mu_{n+1}}\right) = 0, \tag{1.48}$$

$$\partial_{\mu_1}\left(e(*F)^{\mu_1...\mu_{D-n-1}}\right) = 0. \tag{1.49}$$

The field equation (1.48) implies that $*(F-H)_{\mu_1...\mu_{D-n-1}}$ can be locally expressed by a new antisymmetric tensor field $\tilde{B}_{\mu_1...\mu_{D-n-2}}$ of rank $D-n-2$ as

$$*(F-H)_{\mu_1...\mu_{D-n-1}} = (D-n-1)\partial_{[\mu_1}\tilde{B}_{\mu_2...\mu_{D-n-1}]}$$
$$\equiv \tilde{F}_{\mu_1...\mu_{D-n-1}}. \tag{1.50}$$

The field $\tilde{B}_{\mu_1...\mu_{D-n-2}}$ is called a dual field of $B_{\mu_1...\mu_n}$. By using the dual field, the field equation (1.48) and the Bianchi identity (1.49) can be rewritten as

$$\partial_{\mu_1}\left(e(*\tilde{F})^{\mu_1...\mu_{n+1}}\right) = 0, \tag{1.51}$$

$$\partial_{\mu_1}\left(e(\tilde{F}+*H)^{\mu_1...\mu_{D-n-1}}\right) = 0, \tag{1.52}$$

which can be regarded as the Bianchi identity and the field equation of $\tilde{B}_{\mu_1...\mu_{D-n-2}}$. The field equation and the Bianchi identity have been interchanged in (1.48), (1.49) and (1.51), (1.52). The new field equation (1.52) can be derived from a new Lagrangian

$$\tilde{\mathcal{L}} = -\frac{1}{2(D-n-1)!}e\left[\tilde{F}_{\mu_1...\mu_{D-n-1}}\tilde{F}^{\mu_1...\mu_{D-n-1}} + 2\tilde{F}_{\mu_1...\mu_{D-n-1}}(*H)^{\mu_1...\mu_{D-n-1}}\right]$$
$$+ \frac{1}{2(n+1)!}eH_{\mu_1...\mu_{n+1}}H^{\mu_1...\mu_{n+1}}. \tag{1.53}$$

The last term of $\tilde{\mathcal{L}}$ does not depend on $\tilde{F}_{\mu_1...\mu_{D-n-1}}$ and cannot be determined from the field equation. That term is needed by the requirement that the energy–momentum tensor derived from the variation of $\tilde{\mathcal{L}}$ with respect to the metric coincides with that of $\mathcal{L}$. Thus, the antisymmetric tensor field $B_{\mu_1...\mu_n}$ with the Lagrangian (1.47) can be equivalently described by the dual field $\tilde{B}_{\mu_1...\mu_{D-n-2}}$ with the Lagrangian (1.53). Such dualizations of fields will be used in the analysis of dimensional reductions of supergravities in Chap. 6.

There is a more practical method to obtain the dual Lagrangian. Consider a Lagrangian

$$\mathscr{L}' = -\frac{1}{2(n+1)!}\, eF_{\mu_1\ldots\mu_{n+1}} F^{\mu_1\ldots\mu_{n+1}} + \frac{1}{(n+1)!}\, eF_{\mu_1\ldots\mu_{n+1}} H^{\mu_1\ldots\mu_{n+1}}$$

$$+ \frac{1}{(n+1)!(D-n-2)!}\, \varepsilon^{\mu_1\ldots\mu_D} \tilde{B}_{\mu_2\ldots\mu_{D-n-1}} \partial_{\mu_1} F_{\mu_{D-n}\ldots\mu_D}, \qquad (1.54)$$

where $\tilde{B}_{\mu_1\ldots\mu_{D-n-2}}$ and $F_{\mu_1\ldots\mu_{n+1}}$ are independent fields. The variation with respect to $\tilde{B}_{\mu_1\ldots\mu_{D-n-2}}$ gives the Bianchi identity (1.49) for $F_{\mu_1\ldots\mu_{n+1}}$, which implies that $F_{\mu_1\ldots\mu_{n+1}}$ can be written in the form (1.40) by a potential $B_{\mu_1\ldots\mu_n}$. Substituting this back into the Lagrangian (1.54) we obtain the original Lagrangian (1.47), and thus these two Lagrangians $\mathscr{L}'$ and $\mathscr{L}$ are equivalent. On the other hand, the variation of (1.54) with respect to $F_{\mu_1\ldots\mu_{n+1}}$ gives (1.50). Solving this equation for $F_{\mu_1\ldots\mu_{n+1}}$ and substituting it into the Lagrangian (1.54) we obtain the dual Lagrangian (1.53).

1.4.2 Self-dual Field

In $D = 4k+2$ (k is an integer) dimensions we can consider a first order field equation for an antisymmetric tensor field of rank $2k$. The field equation is

$$F^{\mu_1\ldots\mu_{2k+1}} = *F^{\mu_1\ldots\mu_{2k+1}}, \qquad (1.55)$$

i.e., the field strength is equal to its Hodge dual. Such an antisymmetric tensor field is called self-dual. In general even D dimensions the Hodge dual satisfies

$$*^2 F = -(-1)^{\frac{1}{2}D} F. \qquad (1.56)$$

Therefore, the self-duality equation (1.55) can be imposed consistently only in $D = 4k + 2$ dimensions. (In $D = 4k$ dimensions we have a consistent equation if we put an imaginary unit i on the right-hand side of (1.55). In that case, however, F must be a complex field. It can be shown that the self-duality equation for a complex field is equivalent to a Maxwell type field equation for a real field.) For $D = 4k + 2$, acting D_{μ_1} on (1.55), the right-hand side vanishes due to the Bianchi identity (1.44) and we obtain a Maxwell type second order field equation (1.46). Thus, the self-duality equation (1.55) implies the Maxwell type equation. The number of physical degrees of freedom of a self-dual field is half of that of a Maxwell type theory since the field equation is of the first order.

The self-duality equation (1.55) cannot be derived from a simple covariant Lagrangian [5]. One has to introduce an auxiliary field to construct a covariant Lagrangian [7]. We will use a field equation rather than a Lagrangian when we consider a self-dual field.

Self-dual antisymmetric tensor fields appear in supergravities in $D = 10$ and $D = 6$ dimensions. For instance, $D = 10$, $\mathcal{N} = (2, 0)$ supergravity discussed in Sect. 5.4 contains a self-dual antisymmetric tensor field of rank 4.

1.4.3 Massive Chern–Simons Type Theory

In odd D dimensions we can consider a first order field equation for an antisymmetric tensor field of rank $\frac{1}{2}(D-1)$ [12]. The field equation requires that the Hodge dual of the field strength is proportional to the field. Such an equation is some sort of self-duality equation and is called "self-duality in odd dimensions." There is a simple Lagrangian from which the field equation can be derived.

Let us first consider an antisymmetric tensor field of rank $2k-1$ in $D=4k-1$ dimensions. The field equation is

$$* F^{\mu_1 \cdots \mu_{2k-1}} = m B^{\mu_1 \cdots \mu_{2k-1}}, \tag{1.57}$$

where m is a parameter of mass dimension. The field equation (1.57) can be derived from the Lagrangian

$$\mathscr{L} = \frac{m}{2[(2k-1)!]^2} \varepsilon^{\mu_1 \cdots \mu_{4k-1}} B_{\mu_1 \cdots \mu_{2k-1}} \partial_{\mu_{2k}} B_{\mu_{2k+1} \cdots \mu_{4k-1}}$$
$$- \frac{1}{2(2k-1)!} m^2 e B_{\mu_1 \cdots \mu_{2k-1}} B^{\mu_1 \cdots \mu_{2k-1}}, \tag{1.58}$$

which is not invariant under the gauge transformation (1.41). The theory with this Lagrangian is called the massive Chern–Simons type theory. Taking the Hodge dual of (1.57) we find $F^{\mu_1 \cdots \mu_{2k}} = -m * B^{\mu_1 \cdots \mu_{2k}}$. Acting D_{μ_1} on this equation and using (1.57) again on the right-hand side we obtain

$$D_{\mu_1} F^{\mu_1 \cdots \mu_{2k}} - m^2 B^{\mu_2 \cdots \mu_{2k}} = 0. \tag{1.59}$$

This is a Proca type field equation for an antisymmetric tensor field of mass m.

Let us compare the numbers of physical degrees of freedom of a real antisymmetric tensor field of rank $2k-1$ in the massive Chern–Simons type theory, the massive Proca type theory and the massless Maxwell type theory in $D=4k-1$ dimensions. The massless Chern–Simons type theory is a topological field theory [15] and has no local degrees of freedom. The physical degrees of freedom of the Maxwell type theory are transverse components of the field and the number of them is $_{D-2}C_{2k-1}$. The physical degrees of freedom of the Proca type theory are space components of the field since there is no gauge symmetry, and the number of them is $_{D-1}C_{2k-1}$. Finally, the massive Chern–Simons type theory does not have gauge symmetry either but the number of physical degrees of freedom is a half of that of the massive Proca type theory $\frac{1}{2} {}_{D-1}C_{2k-1}$ since the field equation is of the first order. The numbers of physical degrees of freedom in these theories are summarized in Table 1.1. Since $_{D-2}C_{2k-1} = \frac{1}{2} {}_{D-1}C_{2k-1}$ for $D=4k-1$, the massive Chern–Simons type theory has the same number of physical degrees of freedom as the massless Maxwell type theory. As we will see in Sect. 7.3, this fact was used when constructing gauged supergravities in

Table 1.1 The numbers of physical degrees of freedom of an antisymmetric tensor field

Theory	$D = 4k - 1$	$D = 4k + 1$
Massless Maxwell type	$_{D-2}C_{2k-1}$	$_{2D-2}C_{2k}$
Massive Proca type	$_{D-1}C_{2k-1}$	$_{2D-1}C_{2k}$
Massless Chern–Simons type	0	0
Massive Chern–Simons type	$\frac{1}{2}\,_{D-1}C_{2k-1}$	$_{D-1}C_{2k}$

higher dimensions. For instance, $D = 7$, $\mathcal{N} = 4$ gauged supergravity contains third rank antisymmetric tensor fields of the massive Chern–Simons type.

We now turn to the case of an antisymmetric tensor field of rank $2k$ in $D = 4k + 1$ dimensions. The field equation is

$$i * F^{\mu_1 \cdots \mu_{2k}} = m B^{\mu_1 \cdots \mu_{2k}}. \tag{1.60}$$

The imaginary unit i on the left-hand side is necessary by the reason explained below and the field must be a complex field. This field equation can be derived from the Lagrangian

$$\mathcal{L} = \frac{im}{[(2k)!]^2}\, \varepsilon^{\mu_1 \cdots \mu_{4k+1}} B^*_{\mu_1 \cdots \mu_{2k}} \partial_{\mu_{2k+1}} B_{\mu_{2k+2} \cdots \mu_{4k+1}} - \frac{1}{(2k)!}\, m^2 e B^*_{\mu_1 \cdots \mu_{2k}} B^{\mu_1 \cdots \mu_{2k}}. \tag{1.61}$$

As in the case $D = 4k - 1$, we can obtain a Proca type equation from (1.60)

$$D_{\mu_1} F^{\mu_1 \cdots \mu_{2k}} - m^2 B^{\mu_2 \cdots \mu_{2k}} = 0. \tag{1.62}$$

If there was not i on the left-hand side of (1.60), the second term of (1.62) would have an opposite sign and the field would be tachyonic. Actually, the massive Chern–Simons type theory of a complex field is equivalent to the Proca type theory of a real field. By the field equation (1.60) the imaginary part of the field can be expressed in terms of the field strength of the real part. Substituting it back into (1.60) we obtain a Proca type field equation for the real part of the field.

The numbers of physical degrees of freedom of this theory as well as other types of theories for a complex field are summarized in Table 1.1. As in the case $D = 4k - 1$, the massive Chern–Simons type theory in $D = 4k + 1$ has the same number of physical degrees of freedom as the massless Maxwell type theory, and was used in constructing gauged supergravities in higher dimensions. For instance, $D = 5$, $\mathcal{N} = 8$ gauged supergravity contains second rank antisymmetric tensor fields of the massive Chern–Simons type.

References

1. K. Becker, M. Becker, J.H. Schwarz, *String Theory and M-Theory: A Modern Introduction* (Cambridge Univ.Press, Cambridge, 2007)
2. D.Z. Freedman, A. van Proeyen, *Supergravity* (Cambridge Univ.Press, Cambridge, 2012)
3. M.B. Green, J.H. Schwarz and E. Witten, Superstring Theory, Vols. 1, 2 (Cambridge Univ. Press, 1987).
4. C.M. Hull, P.K. Townsend, Unity of superstring dualities. Nucl. Phys. **B438**, 109 (1995). [hep-th/9410167]
5. N. Marcus, J.H. Schwarz, Field theories that have no manifestly Lorentz invariant formulation. Phys. Lett. **B115**, 111 (1982)
6. T. Ortín, *Gravity and Strings* (Cambridge Univ, Press , 2004)
7. P. Pasti, D.P. Sorokin, M. Tonin, Lorentz invariant actions for chiral p-forms. Phys. Rev. **D55**, 6292 (1997). [hep-th/9611100]
8. J. Polchinski, Dirichlet branes and Ramond–Ramond charges. Phys. Rev. Lett. **75**, 4724 (1995). [hep-th/9510017]
9. J. Polchinski, String Theory, Vols. I, II (Cambridge Univ. Press, 1998).
10. A. Salam and E. Sezgin (eds.), Supergravities in Diverse Dimensions, Vols. 1, 2 (North-Holland/World Scientific, 1989).
11. P.K. Townsend, The eleven-dimensional supermembrane revisited. Phys. Lett. **B350**, 184 (1995). [hep-th/9501068]
12. P.K. Townsend, K. Pilch and P. van Nieuwenhuizen, "Selfduality in odd dimensions", Phys. Lett. B136 (1984) 38 [Addendum-ibid. B137 (1984) 443].
13. P. van Nieuwenhuizen, Supergravity. Phys. Rept. **68**, 189 (1981)
14. P.C. West, Introduction to supersymmetry and supergravity, 2nd edition (World Scientific, 1990).
15. E. Witten, Quantum field theory and the Jones polynomial. Commun. Math. Phys. **121**, 351 (1989)
16. E. Witten, String theory dynamics in various dimensions. Nucl. Phys. **B443**, 85 (1995). [hep-th/9503124]
17. B. Zwiebach, *A First Course in String Theory*, 2nd edn. (Cambridge Univ, Press , 2009)

Chapter 2
Supergravities in Four Dimensions

2.1 Superalgebras and Supermultiplets

Supersymmetry is a symmetry between bosons and fermions. In supersymmetric theories bosons and fermions belong to supermultiplets and are related by supertransformations. Supertransformations together with spacetime transformations such as the Poincaré transformations form a superalgebra. There are various kinds of superalgebras depending on the spacetime dimension, the spacetime symmetry and the number of supersymmetries. In this section we discuss supermultiplets of the $D = 4$ super Poincaré algebra with the smallest number of supersymmetries.

The super Poincaré algebra consists of the generators of supertransformations (supercharges) Q_α and those of the Poincaré algebra, i.e., the translation generators P_μ and the Lorentz generators $M_{\mu\nu}(= -M_{\nu\mu})$. (We use $\alpha, \beta, \ldots = 1, 2, 3, 4$ for spinor indices.) When it contains only one Majorana spinor supercharge, it is called the $\mathcal{N} = 1$ super Poincaré algebra, which we consider first. A Majorana spinor ψ is a spinor satisfying the Majorana condition $\psi = \psi^c$, where ψ^c is the charge conjugation of ψ defined by

$$\psi^c = C\bar{\psi}^T. \tag{2.1}$$

Here, the charge conjugation matrix C is a 4×4 matrix satisfying

$$C^{-1}\gamma^\mu C = -\gamma^{\mu T}, \quad C^T = -C, \quad C^\dagger C = 1. \tag{2.2}$$

The (anti)commutation relations of the $\mathcal{N} = 1$ super Poincaré algebra are

$$[M_{\mu\nu}, M_{\rho\sigma}] = -i\eta_{\nu\rho}M_{\mu\sigma} + i\eta_{\nu\sigma}M_{\mu\rho} + i\eta_{\mu\rho}M_{\nu\sigma} - i\eta_{\mu\sigma}M_{\nu\rho},$$
$$[M_{\mu\nu}, P_\rho] = -i\eta_{\nu\rho}P_\mu + i\eta_{\mu\rho}P_\nu, \quad [P_\mu, P_\nu] = 0,$$
$$[M_{\mu\nu}, Q_\alpha] = \frac{1}{2}i(\gamma_{\mu\nu})_\alpha{}^\beta Q_\beta, \quad [P_\mu, Q_\alpha] = 0,$$
$$\{Q_\alpha, \bar{Q}^\beta\} = -2i(\gamma^\mu)_\alpha{}^\beta P_\mu, \tag{2.3}$$

Y. Tanii, *Introduction to Supergravity*, SpringerBriefs in Mathematical Physics, DOI: 10.1007/978-4-431-54828-7_2, © The Author(s) 2014

where $\eta_{\mu\nu} = \mathrm{diag}(-1, +1, +1, +1)$ is the metric of flat Minkowski spacetime and $\gamma^{\mu\nu} = \gamma^{[\mu}\gamma^{\nu]}$ is the antisymmetrized product of gamma matrices defined in Appendix B. The components of the supercharge Q_α are fermionic generators and satisfy anticommutation relations rather than commutation relations.

In supersymmetric theories a certain set of particle states with different spins form a multiplet of the superalgebra called a supermultiplet. We can find possible supermultiplets by studying irreducible representations of the super Poincaré algebra for one particle states. From the fifth commutation relation of (2.3) we see that all the states generated by acting Q_α on a state in a supermultiplet have the same eigenvalue p_μ of the translation generator P_μ. Hence, all such states have the same mass $m = \sqrt{-p_\mu p^\mu}$. To proceed we choose a representation of the gamma matrices

$$\gamma^0 = -\mathrm{i}\begin{pmatrix} 0 & 1 \\ 1 & 0 \end{pmatrix}, \quad \gamma^i = -\mathrm{i}\begin{pmatrix} 0 & \sigma_i \\ -\sigma_i & 0 \end{pmatrix} \quad (i = 1, 2, 3), \tag{2.4}$$

where σ_i are the 2×2 Pauli matrices

$$\sigma_1 = \begin{pmatrix} 0 & 1 \\ 1 & 0 \end{pmatrix}, \quad \sigma_2 = \begin{pmatrix} 0 & -\mathrm{i} \\ \mathrm{i} & 0 \end{pmatrix}, \quad \sigma_3 = \begin{pmatrix} 1 & 0 \\ 0 & -1 \end{pmatrix}. \tag{2.5}$$

In this representation the chirality matrix $\gamma_5 = \mathrm{i}\,\gamma_0\gamma_1\gamma_2\gamma_3$ is diagonal. The charge conjugation matrix C and the Majorana spinor supercharge Q can be written as

$$C = \begin{pmatrix} \mathrm{i}\sigma_2 & 0 \\ 0 & -\mathrm{i}\sigma_2 \end{pmatrix}, \quad Q = \begin{pmatrix} \mathrm{i}\sigma_2 Q^{\dagger T} \\ Q \end{pmatrix}. \tag{2.6}$$

We can choose the lower two components Q_3, Q_4 as independent components of the supercharge.

Let us first consider supermultiplets of massless one particle states. In this case we can choose a Lorentz frame in which the momentum eigenvalue is $p^\mu = (E, 0, 0, E)$ $(E > 0)$. Then, only non-vanishing anticommutator of Q_α $(\alpha = 3, 4)$ in (2.3) is

$$\{Q_3, (Q_3)^\dagger\} = 4E. \tag{2.7}$$

Since Q_4 anticommutes with all the components of Q including Q_4 itself, we can assume $Q_4 = 0$. If we define $b = (4E)^{-\frac{1}{2}}Q_3$, $b^\dagger = (4E)^{-\frac{1}{2}}Q_3^\dagger$, they satisfy the anticommutation relations of creation and annihilation operators of a fermion

$$\{b, b^\dagger\} = 1, \quad \{b, b\} = 0, \quad \{b^\dagger, b^\dagger\} = 0. \tag{2.8}$$

Therefore, their representation space consists of two states

$$|h_0\rangle, \quad b^\dagger|h_0\rangle, \tag{2.9}$$

where $|h_0\rangle$ is a state with helicity h_0 satisfying $b\,|h_0\rangle = 0$. From the fourth commutation relation in (2.3) we find $[M_{12}, b^\dagger] = \frac{1}{2}b^\dagger$, which implies that $b^\dagger$ has helicity $\frac{1}{2}$. Therefore, the states (2.9) have helicities $(h_0, h_0 + \frac{1}{2})$. According to the CPT theorem, if a quantum field theory contains a state with helicity h, then it also contains a state with helicity $-h$. Therefore, supermultiplets realized by a quantum field theory are

$$(h_0,\ h_0 + \tfrac{1}{2})\ \oplus\ (-h_0 - \tfrac{1}{2},\ -h_0) \quad (h_0 = 0,\ \tfrac{1}{2},\ 1, \ldots). \tag{2.10}$$

Massless supermultiplets often used in particle physics are

$$\text{chiral multiplet:}\quad (0,\ \tfrac{1}{2}) \oplus (-\tfrac{1}{2},\ 0),$$
$$\text{massless vector multiplet:}\quad (\tfrac{1}{2},\ 1) \oplus (-1,\ -\tfrac{1}{2}),$$
$$\text{supergravity multiplet:}\quad (\tfrac{3}{2},\ 2) \oplus (-2,\ -\tfrac{3}{2}). \tag{2.11}$$

For massive one particle states with a mass m we can choose a Lorentz frame in which the momentum eigenvalue is $p^\mu = (m, 0, 0, 0)$. Then, the anticommutator of the supercharge Q_α ($\alpha = 3,\ 4$) in (2.3) becomes

$$\{Q_\alpha, (Q_\beta)^\dagger\} = 2m\delta_{\alpha\beta}. \tag{2.12}$$

This implies that $b_\alpha = (2m)^{-\frac{1}{2}} Q_\alpha$ and $b_\alpha^\dagger = (2m)^{-\frac{1}{2}} Q_\alpha^\dagger$ are two sets of creation and annihilation operators of fermions. Therefore, their representation space consists of

$$|s_0\rangle,\quad b_\alpha^\dagger\,|s_0\rangle,\quad b_3^\dagger b_4^\dagger\,|s_0\rangle. \tag{2.13}$$

The state $|s_0\rangle$ satisfies $b_\alpha\,|s_0\rangle = 0$ and collectively represents the $2s_0 + 1$ states of spin s_0. Since $b_\alpha^\dagger$ has spin $\frac{1}{2}$ as can be seen from the fourth commutation relation in (2.3), the states (2.13) have spins

$$(s_0 - \tfrac{1}{2},\ s_0,\ s_0,\ s_0 + \tfrac{1}{2}) \quad (s_0 = 0,\ \tfrac{1}{2},\ 1, \ldots), \tag{2.14}$$

where the $s_0 - \frac{1}{2}$ state is absent for $s_0 = 0$. Massive supermultiplets often used are

$$\text{chiral multiplet:}\quad (0,\ 0,\ \tfrac{1}{2}),$$
$$\text{massive vector multiplet:}\quad (0,\ \tfrac{1}{2},\ \tfrac{1}{2},\ 1). \tag{2.15}$$

We see that the numbers of bosonic states and fermionic states in a supermultiplet are the same for both of massless and massive supermultiplets.

2.2 Supersymmetric Field Theories

To construct a supersymmetric field theory one introduces supermultiplets of fields. In globally supersymmetric theories one usually uses two kinds of supermultiplets:

$$\text{vector multiplet: } (A_\mu^I, \chi^I, D^I) \quad (I = 1, 2, \ldots, \dim G),$$
$$\text{chiral multiplet: } (\phi_i, \psi_{-i}, F_i) \quad (i = 1, 2, \ldots, n). \tag{2.16}$$

The vector multiplet consists of vector fields $A_\mu^I(x)$, Majorana spinor fields $\chi^I(x)$ and real scalar fields $D^I(x)$, which all belong to the adjoint representation of a gauge group G. The chiral multiplet consists of complex scalar fields $\phi_i(x)$, $F_i(x)$ and Weyl spinor fields with negative chirality $\psi_{-i}(x)$ ($\gamma_5 \psi_{-i} = -\psi_{-i}$), which all belong to a certain n-dimensional representation of G. The fields D^I and F_i are auxiliary fields, which do not have physical degrees of freedom, and can be expressed in terms of other fields by their field equations. To construct a supersymmetric Lagrangian for these fields it is useful to use the superfield method, which we do not discuss here (see, e.g., [16]).

As an example of supersymmetric field theories let us consider a theory consisting of a chiral multiplet (ϕ, ψ_-, F). A supersymmetric Lagrangian is

$$\mathscr{L} = - \partial_\mu \phi^* \partial^\mu \phi - \bar{\psi}_- \gamma^\mu \partial_\mu \psi_- + F^* F + W'(\phi) F + (W'(\phi))^* F^*$$
$$- \frac{1}{2} W''(\phi) \bar{\psi}_+ \psi_- - \frac{1}{2} (W''(\phi))^* \bar{\psi}_- \psi_+, \tag{2.17}$$

where the superpotential $W(\phi)$ is a holomorphic function of the complex scalar field ϕ, and $\psi_+ = (\psi_-)^c$ is the charge conjugation of ψ_- and has positive chirality ($\gamma_5 \psi_+ = +\psi_+$). This Lagrangian is invariant up to total divergences under the supertransformation

$$\delta_Q \phi = \frac{1}{2} \bar{\varepsilon}_+ \psi_-, \quad \delta_Q \psi_- = \frac{1}{2} \gamma^\mu \varepsilon_+ \partial_\mu \phi + \frac{1}{2} F \varepsilon_-, \quad \delta_Q F = \frac{1}{2} \bar{\varepsilon}_- \gamma^\mu \partial_\mu \psi_-. \tag{2.18}$$

Therefore, the action obtained by integrating the Lagrangian over spacetime is invariant. The transformation parameter $\varepsilon = \varepsilon_+ + \varepsilon_-$ is a constant Majorana spinor and $\varepsilon_\pm$ are its projections on the chirality eigenstates. Since the transformation parameter is fermionic, the supertransformation exchanges the bosonic fields ϕ, F and the fermionic field ψ. For all the fields the commutator of two supertransformations with parameters ε_1 and ε_2 becomes

$$[\delta_Q(\varepsilon_1), \delta_Q(\varepsilon_2)] = \delta_P(\xi), \quad \xi^\mu = \frac{1}{4} \bar{\varepsilon}_2 \gamma^\mu \varepsilon_1. \tag{2.19}$$

where $\delta_P(\xi)$ is the infinitesimal translation with a parameter ξ^μ. This commutation relation corresponds to the last anticommutation relation in the super Poincaré algebra (2.3).

The field equation of the auxiliary field F derived from the Lagrangian (2.17) is algebraic and can be used to express it in terms of the scalar field ϕ as $F^* = -W'(\phi)$. Substituting this back into (2.17) we obtain a Lagrangian without the auxiliary field. For instance, if we choose the superpotential $W(\phi) = \frac{1}{2}m\phi^2 + \frac{1}{6}\lambda\phi^3$, we obtain

$$\mathcal{L} = -\partial_\mu\phi^*\partial^\mu\phi - \bar{\psi}_-\gamma^\mu\partial_\mu\psi_- - \left|m\phi + \frac{1}{2}\lambda\phi^2\right|^2$$
$$- \frac{1}{2}(m+\lambda\phi)\bar{\psi}_+\psi_- - \frac{1}{2}(m+\lambda\phi^*)\bar{\psi}_-\psi_+. \tag{2.20}$$

The fields ϕ and ψ_- represent two spin 0 bosons and a spin $\frac{1}{2}$ fermion with the same mass m. They form a chiral multiplet $(0, 0, \frac{1}{2})$ in (2.15). The coupling constants of various interaction terms in (2.20) are related by supersymmetry. For instance, the coupling constant of the $|\phi|^4$ coupling and that of the Yukawa coupling $\phi\bar{\psi}_+\psi_-$ are given by using the same λ. Substituting $F^* = -W'(\phi)$ into the first two equations in (2.18) we obtain the supertransformation of ϕ and ψ_- without the auxiliary field. The Lagrangian with the auxiliary field eliminated is still invariant up to total divergences under this transformation. However, an extra term proportional to the field equation of ψ_- appears on the right-hand side of the commutation relation (2.19) for ψ_-. Thus, when the auxiliary field is eliminated, the commutator algebra closes only on-shell, i.e., only if the field equation is used.

2.3 $\mathcal{N} = 1$ Poincaré Supergravity

Supergravity is a field theory which has local supersymmetry. Since the transformation parameter of supersymmetry is a spinor ε_α, the gauge field of local supersymmetry should be $\psi_{\mu\alpha}(x)$ with a vector index μ and a spinor index α. The transformation law of this gauge field is $\delta_Q\psi_{\mu\alpha} = \partial_\mu\varepsilon_\alpha + \cdots$, where the transformation parameter $\varepsilon_\alpha(x)$ is an arbitrary function of spacetime coordinates x^μ. Such a field $\psi_{\mu\alpha}(x)$ is the Rarita–Schwinger field representing a spin $\frac{3}{2}$ fermion (gravitino). Furthermore, we need another gauge field. In globally supersymmetric theories the commutator of two supertransformations generates a translation as in (2.19). Therefore, we expect that the gauging of supersymmetry leads to the gauging of translation. Since the local translation is the general coordinate transformation, we also need the gravitational field $e_\mu{}^a(x)$ as a gauge field. To summarize, supergravity is a theory which is invariant under the local supersymmetry transformation as well as the general coordinate transformation. It contains the gravitational field $e_\mu{}^a(x)$ and the Rarita–Schwinger field $\psi_{\mu\alpha}(x)$. As we saw in Sect. 2.1, the supergravity multiplet for the $\mathcal{N} = 1$ super Poincaré algebra consists of the states with helicities $(\frac{3}{2}, 2) \oplus (-2, -\frac{3}{2})$, which correspond to a pair of fields $(e_\mu{}^a(x), \psi_{\mu\alpha}(x))$. Hence, we expect that there exists a supergravity theory which contains these two fields. Such a theory was indeed constructed in [2, 9, 10] and is called $\mathcal{N} = 1$ Poincaré supergravity.

The field content of $\mathcal{N} = 1$ Poincaré supergravity is a vierbein $e_\mu{}^a(x)$ and a Majorana Rarita–Schwinger field $\psi_\mu(x)$. As discussed in Chap. 1 the vierbein is related to the metric as $g_{\mu\nu} = e_\mu{}^a e_\nu{}^b \eta_{ab}$, where $\eta_{ab} = \mathrm{diag}(-1, +1, +1, +1)$ is the flat Minkowski metric. The Rarita–Schwinger field satisfies the Majorana condition $\psi_\mu^c = \psi_\mu$. The Lagrangian consists of the Einstein term and the Rarita–Schwinger term as

$$\mathscr{L} = e\hat{R} - \frac{1}{2}e\bar{\psi}_\mu \gamma^{\mu\nu\rho}\hat{D}_\nu\psi_\rho, \tag{2.21}$$

where $e = \det e_\mu{}^a$, $\gamma^\mu = \gamma^a e_a{}^\mu$, and $\gamma^{\mu\nu\rho} = \gamma^{[\mu}\gamma^\nu\gamma^{\rho]}$ is the antisymmetrized product of gamma matrices defined in Appendix B. The curvature and the covariant derivative are defined by

$$\hat{R} = e_a{}^\mu e_b{}^\nu \hat{R}_{\mu\nu}{}^{ab},$$
$$\hat{R}_{\mu\nu}{}^{ab} = \partial_\mu\hat{\omega}_\nu{}^{ab} - \partial_\nu\hat{\omega}_\mu{}^{ab} + \hat{\omega}_\mu{}^a{}_c\hat{\omega}_\nu{}^{cb} - \hat{\omega}_\nu{}^a{}_c\hat{\omega}_\mu{}^{cb},$$
$$\hat{D}_{[\nu}\psi_{\rho]} = \left(\partial_{[\nu} + \frac{1}{4}\hat{\omega}_{[\nu}{}^{ab}\gamma_{ab}\right)\psi_{\rho]}. \tag{2.22}$$

The spin connection $\hat{\omega}_\mu{}^{ab}$ used here is given by

$$\hat{\omega}_{\mu ab} = \omega_{\mu ab} + \frac{1}{8}\left(\bar{\psi}_a\gamma_\mu\psi_b + \bar{\psi}_\mu\gamma_a\psi_b - \bar{\psi}_\mu\gamma_b\psi_a\right), \tag{2.23}$$

where $\omega_{\mu ab}$ is the spin connection without torsion given in (1.21). The spin connection (2.23) has a torsion depending on the Rarita–Schwinger field:

$$\hat{D}_\mu e_\nu{}^a - \hat{D}_\nu e_\mu{}^a = \frac{1}{4}\bar{\psi}_\mu\gamma^a\psi_\nu. \tag{2.24}$$

If one wishes, one can rewrite the Lagrangian using the torsionless spin connection $\omega_{\mu ab}$ as

$$\mathscr{L} = eR - \frac{1}{2}e\bar{\psi}_\mu\gamma^{\mu\nu\rho}D_\nu\psi_\rho + \text{(four-fermi terms)}, \tag{2.25}$$

where R and D_ν are defined by using $\omega_{\mu ab}$. Explicit four-fermi terms have appeared.

The Lagrangian (2.21) is invariant up to total divergences under the general coordinate transformation

$$\delta_G(\xi)e_\mu{}^a = \xi^\nu\partial_\nu e_\mu{}^a + \partial_\mu\xi^\nu e_\nu{}^a, \quad \delta_G(\xi)\psi_\mu = \xi^\nu\partial_\nu\psi_\mu + \partial_\mu\xi^\nu\psi_\nu, \tag{2.26}$$

the local Lorentz transformation

$$\delta_L(\lambda)e_\mu{}^a = -\lambda^a{}_b e_\mu{}^b, \quad \delta_L(\lambda)\psi_\mu = -\frac{1}{4}\lambda^{ab}\gamma_{ab}\psi_\mu \tag{2.27}$$

and the $\mathcal{N} = 1$ local supertransformation

$$\delta_Q(\varepsilon)e_\mu{}^a = \frac{1}{4}\bar{\varepsilon}\gamma^a\psi_\mu, \quad \delta_Q(\varepsilon)\psi_\mu = \hat{D}_\mu\varepsilon, \tag{2.28}$$

where the transformation parameters $\xi^\mu(x)$, $\lambda_{ab}(x)$ $(\lambda_{ab} = -\lambda_{ba})$ and $\varepsilon_\alpha(x)$ $(\varepsilon^c = \varepsilon)$ are arbitrary infinitesimal functions of spacetime coordinates x^μ. The invariance under the first two bosonic transformations is manifest. The invariance under the local supertransformation is shown in the next section. From (2.28) we can compute the local supertransformation of the spin connection $\hat{\omega}_{\mu ab}$ in (2.23). We see that terms containing derivatives of the transformation parameter $\partial_\mu\varepsilon$ appear from $\delta_Q\omega_{\mu ab}$ but are canceled by the variation of the ψ_μ bilinear terms (see (2.40) below). In general, quantities whose local supertransformation does not contain $\partial_\mu\varepsilon$ are called supercovariant. We often put ^ on supercovariant quantities.

The above local transformations satisfy the closed commutation relations

$$[\delta_G(\xi_1), \delta_G(\xi_2)] = \delta_G(\xi_2 \cdot \partial\xi_1 - \xi_1 \cdot \partial\xi_2),$$
$$[\delta_L(\lambda_1), \delta_L(\lambda_2)] = \delta_L([\lambda_1, \lambda_2]),$$
$$[\delta_G(\xi), \delta_L(\lambda)] = \delta_L(-\xi \cdot \partial\lambda),$$
$$[\delta_G(\xi), \delta_Q(\varepsilon)] = \delta_Q(-\xi \cdot \partial\varepsilon),$$
$$[\delta_L(\lambda), \delta_Q(\varepsilon)] = \delta_Q(\tfrac{1}{4}\lambda^{ab}\gamma_{ab}\varepsilon),$$
$$[\delta_Q(\varepsilon_1), \delta_Q(\varepsilon_2)] = \delta_G(\xi) + \delta_L(\lambda) + \delta_Q(\varepsilon), \tag{2.29}$$

where the transformation parameters on the right-hand side of the last commutation relation are

$$\xi^\mu = \frac{1}{4}\bar{\varepsilon}_2\gamma^\mu\varepsilon_1, \quad \lambda_{ab} = -\xi^\mu\hat{\omega}_{\mu ab}, \quad \varepsilon = -\xi^\mu\psi_\mu. \tag{2.30}$$

We see that the commutator of two local supertransformations generates a general coordinate transformation as we expected at the beginning of this section. The commutation relations (2.29) except the last one can be easily shown. The last commutation relation is shown in the next section. To obtain the last commutation relation we have to use the Rarita–Schwinger field equation derived from the Lagrangian (2.21). In this sense the commutator algebra closes only on-shell. In the present theory it is possible to close the commutator algebra off-shell by introducing an appropriate set of auxiliary fields, which have no dynamical degrees of freedom [5, 12, 13]. A formulation with an off-shell algebra is more convenient, although not indispensable, when one fixes a gauge of the local symmetries and when one couples matter supermultiplets. For general supergravities (those with highly extended supersymmetry and/or in higher dimensions) such an off-shell formulation is not known.

This theory also has a global symmetry. The Lagrangian (2.21) is invariant under the global chiral $U(1)$ transformation

$$\delta e_\mu{}^a = 0, \qquad \delta\psi_\mu = i\Lambda\gamma_5\psi_\mu, \qquad (2.31)$$

where Λ is a constant infinitesimal transformation parameter. The transformation of ψ_μ is consistent with the Majorana condition on ψ_μ.

The field equations derived from the Lagrangian (2.21) have a Minkowski space-time solution $e_\mu{}^a = \delta_\mu^a$, $\psi_\mu = 0$. This solution is preserved by the super Poincaré transformations corresponding to the local symmetry transformations with the parameters $\xi^\mu(x) = a^\mu{}_\nu x^\nu + b^\mu$, $\lambda^\mu{}_\nu(x) = a^\mu{}_\nu$, $\varepsilon(x) = \varepsilon$, where $a^\mu{}_\nu$, b^μ and ε are constant. Dynamics of small fluctuations of the fields around this background is subject to the symmetry under these transformations. Substituting this solution and these transformation parameters into (2.29) we find the commutation relations of the super Poincaré algebra (2.3). In general, a supergravity which has a Minkowski spacetime solution is called the Poincaré supergravity.

One can couple matter supermultiplets to the supergravity multiplet $(e_\mu{}^a, \psi_\mu)$. Possible matter multiplets are the chiral multiplet (ϕ_i, ψ_{-i}) and the vector multiplet (A^I_μ, χ^I), which we discussed in Sect. 2.2. For details of matter couplings see [8].

2.4 Local Supersymmetry of $\mathcal{N} = 1$ Poincaré Supergravity

In this section we show the invariance of the action and the commutator algebra of $\mathcal{N} = 1$ Poincaré supergravity discussed in the previous section. We use the identities for spinors and gamma matrices given in Appendix B.

2.4.1 Invariance of the Action

The Lagrangian (2.21) consists of two terms:

$$\mathcal{L} = \mathcal{L}_E + \mathcal{L}_{RS},$$
$$\mathcal{L}_E = e\, e_a{}^\mu e_b{}^\nu \hat{R}_{\mu\nu}{}^{ab} = -\frac{1}{4}\varepsilon^{\mu\nu\rho\sigma}\varepsilon_{abcd}e_\rho{}^c e_\sigma{}^d \hat{R}_{\mu\nu}{}^{ab},$$
$$\mathcal{L}_{RS} = -\frac{1}{2}e e_a{}^\mu e_b{}^\nu e_c{}^\rho \bar{\psi}_\mu \gamma^{abc} \hat{D}_\nu \psi_\rho = \frac{1}{2}i\varepsilon^{\mu\nu\rho\sigma}\bar{\psi}_\mu \gamma_\nu \gamma_5 \hat{D}_\rho \psi_\sigma, \qquad (2.32)$$

where $\varepsilon^{\mu\nu\rho\sigma}$ and ε_{abcd} are the totally antisymmetric Levi-Civita symbols with components $\varepsilon^{0123} = +1$ and $\varepsilon_{0123} = -1$ (see Appendix A), and $\gamma_5 = i\gamma_0\gamma_1\gamma_2\gamma_3$ is the chirality matrix. The Riemann tensor $\hat{R}_{\mu\nu}{}^{ab}$ and the covariant derivative $\hat{D}_\mu$ depend on the fields $e_\mu{}^a$, ψ_μ only through the spin connection $\hat{\omega}_{\mu ab}$. The following observation is useful to show the invariance of the action. If we treat $\hat{\omega}_{\mu ab}$ as an independent field, the variation of the action with respective to it vanishes:

$$\frac{\delta}{\delta\hat{\omega}_{\mu ab}}\int d^4x\, \mathcal{L}(e, \psi, \hat{\omega}) = 0 \qquad (2.33)$$

when (2.23) is substituted into (2.33) after the variation. To show this, it is convenient to use the second forms of $\mathscr{L}_E$ and $\mathscr{L}_{RS}$ in (2.32). Therefore, when we compute the supertransformation of the Lagrangian, we need not consider the variation of the spin connection.

Let us compute the variation of the Lagrangian (2.32) under the local supertransformation (2.28). Using the first form of $\mathscr{L}_E$ in (2.32) we find

$$
\begin{aligned}
\delta_Q \mathscr{L}_E &= \delta_Q (e\, e_a{}^\mu e_b{}^\nu) \hat{R}_{\mu\nu}{}^{ab} \\
&= -\frac{1}{2} e \bar{\varepsilon} \gamma^\mu \psi_a \left(e_b{}^\nu \hat{R}_{\mu\nu}{}^{ab} - \frac{1}{2} e_\mu{}^a \hat{R} \right),
\end{aligned}
\tag{2.34}
$$

while using the second form of $\mathscr{L}_{RS}$ in (2.32) we find

$$
\begin{aligned}
\delta_Q \mathscr{L}_{RS} = \frac{1}{2} i \varepsilon^{\mu\nu\rho\sigma} \Big(&\hat{D}_\mu \bar{\varepsilon} \gamma_\nu \gamma_5 \hat{D}_\rho \psi_\sigma + \bar{\psi}_\mu \gamma_\nu \gamma_5 \hat{D}_\rho \hat{D}_\sigma \varepsilon \\
&+ \delta_Q e_\nu{}^a \, \bar{\psi}_\mu \gamma_a \gamma_5 \hat{D}_\rho \psi_\sigma \Big).
\end{aligned}
\tag{2.35}
$$

By partial integration, the first term of (2.35) becomes

$$
-\frac{1}{2} i \varepsilon^{\mu\nu\rho\sigma} \bar{\varepsilon} \gamma_\nu \gamma_5 \hat{D}_\mu \hat{D}_\rho \psi_\sigma - \frac{1}{2} i \varepsilon^{\mu\nu\rho\sigma} \hat{D}_\mu e_\nu{}^a \, \bar{\varepsilon} \gamma_a \gamma_5 \hat{D}_\rho \psi_\sigma
\tag{2.36}
$$

up to total divergences. By using (2.24), the Fierz identity (B.15) and the symmetry properties (B.13) we find that the second term of (2.36) cancels the third term of (2.35). Then, (2.35) becomes

$$
\begin{aligned}
\delta_Q \mathscr{L}_{RS} &= -\frac{1}{4} i \varepsilon^{\mu\nu\rho\sigma} \bar{\varepsilon} \gamma_\nu \gamma_5 [\hat{D}_\mu, \hat{D}_\rho] \psi_\sigma + \frac{1}{4} i \varepsilon^{\mu\nu\rho\sigma} \bar{\psi}_\mu \gamma_\nu \gamma_5 [\hat{D}_\rho, \hat{D}_\sigma] \varepsilon \\
&= \frac{1}{2} e \bar{\varepsilon} \gamma^\mu \psi_a \left(e_b{}^\nu \hat{R}_{\mu\nu}{}^{ab} - \frac{1}{2} e_\mu{}^a \hat{R} \right)
\end{aligned}
\tag{2.37}
$$

up to total divergences, where in the last equality we have used (1.30) and (B.13). Thus, the variations of $\mathscr{L}_E$ and $\mathscr{L}_{RS}$ cancel each other and the total Lagrangian (2.21) is invariant under the local supertransformation (2.28) up to total divergences.

2.4.2 Commutator Algebra

Next let us show the commutation relation of two local supertransformations in (2.29). We first consider the commutator acting on the vierbein

$$[\delta_Q(\varepsilon_1), \delta_Q(\varepsilon_2)]e_\mu{}^a = \delta_Q(\varepsilon_1)\left(\frac{1}{4}\bar{\varepsilon}_2\gamma^a\psi_\mu\right) - (1 \leftrightarrow 2)$$

$$= \frac{1}{4}\bar{\varepsilon}_2\gamma^a\hat{D}_\mu\varepsilon_1 - \frac{1}{4}\bar{\varepsilon}_1\gamma^a\hat{D}_\mu\varepsilon_2$$

$$= \frac{1}{4}\hat{D}_\mu\left(\bar{\varepsilon}_2\gamma^a\varepsilon_1\right), \tag{2.38}$$

where we have used (B.13). Then, we obtain

$$[\delta_Q(\varepsilon_1), \delta_Q(\varepsilon_2)]e_\mu{}^a = \hat{D}_\mu\left(\xi^\nu e_\nu{}^a\right)$$

$$= \partial_\mu\xi^\nu e_\nu{}^a + \xi^\nu\hat{D}_\nu e_\mu{}^a + \xi^\nu\left(\hat{D}_\mu e_\nu{}^a - \hat{D}_\nu e_\mu{}^a\right)$$

$$= \partial_\mu\xi^\nu e_\nu{}^a + \xi^\nu\partial_\nu e_\mu{}^a + \xi^\nu\hat{\omega}_\nu{}^a{}_b e_\mu{}^b - \frac{1}{4}\xi^\nu\bar{\psi}_\nu\gamma^a\psi_\mu$$

$$= \left[\delta_G(\xi) + \delta_L(-\xi\cdot\hat{\omega}) + \delta_Q(-\xi\cdot\psi)\right]e_\mu{}^a, \tag{2.39}$$

where $\xi^\nu = \frac{1}{4}\bar{\varepsilon}_2\gamma^\nu\varepsilon_1$ and we have used (2.24). This shows the last commutation relation in (2.29) for $e_\mu{}^a$. Similarly, we can compute the commutation relation for ψ_μ. We need the supertransformation of the spin connection $\hat{\omega}_{\mu ab}$, which can be obtained by applying δ_Q on both sides of (2.24) as

$$\delta_Q(\varepsilon)\hat{\omega}_{\mu ab} = -\frac{1}{8}\left(\bar{\varepsilon}\gamma_\mu\psi_{ab} - \bar{\varepsilon}\gamma_a\psi_{b\mu} + \bar{\varepsilon}\gamma_b\psi_{a\mu}\right), \tag{2.40}$$

where $\psi_{\mu\nu} = \hat{D}_\mu\psi_\nu - \hat{D}_\nu\psi_\mu$. By using (2.40), (B.15) and (B.13) we obtain

$$[\delta_Q(\varepsilon_1), \delta_Q(\varepsilon_2)]\psi_\mu = \left[\delta_G(\xi) + \delta_L(-\xi\cdot\hat{\omega}) + \delta_Q(-\xi\cdot\psi)\right]\psi_\mu$$

$$+ \frac{1}{128}\bar{\varepsilon}_2\gamma^{ab}\varepsilon_1\left(2\gamma_{ab\mu\nu}\mathscr{R}^\nu - \gamma_{ab}\mathscr{R}_\mu - 2e_{\mu a}\mathscr{R}_b\right)$$

$$- \frac{1}{16}\xi^\nu\left(\gamma_\nu\mathscr{R}_\mu + 2\gamma_{\mu\nu\lambda}\mathscr{R}^\lambda\right), \tag{2.41}$$

where $\mathscr{R}^\nu = \gamma^{\nu\rho\sigma}\psi_{\rho\sigma}$. The field equation of the Rarita–Schwinger field is $\mathscr{R}^\nu = 0$. Therefore, the commutator algebra closes on-shell.

2.5 $\mathscr{N} = 1$ Anti de Sitter Supergravity

We can construct a supergravity with a cosmological term [14]. The Lagrangian is

$$\mathscr{L} = e\hat{R} - \frac{1}{2}e\bar{\psi}_\mu\gamma^{\mu\nu\rho}\hat{D}_\nu\psi_\rho + 6m^2e + \frac{1}{2}me\bar{\psi}_\mu\gamma^{\mu\nu}\psi_\nu, \tag{2.42}$$

where m is a real constant parameter. The third term is the cosmological term. Comparing with (1.2) in Chap. 1 we see that the cosmological constant is negative $\Lambda = -3m^2$. The last term is a mass term of the Rarita–Schwinger field and is proportional to the parameter m appearing in the cosmological term. When $m = 0$, this theory reduces to the Poincaré supergravity in Sect. 2.3. A positive cosmological constant corresponds to an imaginary m, for which the Rarita–Schwinger mass term is not real. Therefore, a positive cosmological constant is not allowed.

The Lagrangian (2.42) is invariant up to total divergences under the general coordinate transformation (2.26), the local Lorentz transformation (2.27) and the $\mathcal{N} = 1$ local supertransformation

$$\delta_Q e_\mu{}^a = \frac{1}{4}\bar{\varepsilon}\gamma^a\psi_\mu, \qquad \delta_Q \psi_\mu = \hat{D}_\mu\varepsilon + \frac{1}{2}m\gamma_\mu\varepsilon. \tag{2.43}$$

The term proportional to m has been added to $\delta_Q \psi_\mu$. The commutator algebra of the local transformations has the same form as (2.29) for $m = 0$ except that the parameter of the local Lorentz transformation in (2.30) is replaced by $\lambda_{ab} = -\xi^\mu\hat{\omega}_{\mu ab} - \frac{1}{4}m\bar{\varepsilon}_2\gamma_{ab}\varepsilon_1$. To obtain the closed commutator algebra we have to use the Rarita–Schwinger field equation derived from (2.42). Due to the mass term of the Rarita–Schwinger field the Lagrangian is no longer invariant under the global $U(1)$ transformation (2.31).

Let us consider a classical solution of the field equations derived from the Lagrangian (2.42). When $\psi_\mu = 0$, the Rarita–Schwinger field equation is automatically satisfied and the gravitational field equation becomes

$$R_{\mu\nu} = -3m^2 g_{\mu\nu}. \tag{2.44}$$

Minkowski spacetime $e_\mu{}^a = \delta_\mu^a$ has $R_{\mu\nu} = 0$ and therefore does not satisfy this equation. A solution of this equation is anti de Sitter (AdS) spacetime (see, e.g., [8] for details of AdS spacetime). The Riemann tensor of AdS spacetime can be expressed by using the metric as

$$R_{\mu\nu\rho\sigma} = -m^2 \left(g_{\mu\rho}g_{\nu\sigma} - g_{\mu\sigma}g_{\nu\rho}\right) \tag{2.45}$$

and the Ricci tensor satisfies (2.44). The parameter m is called the inverse radius of AdS spacetime. Supergravities which have an AdS spacetime solution are called anti de Sitter (AdS) supergravities.

AdS spacetime has a large isometry $SO(2, 3)$ in the same way as Minkowski spacetime has the isometry of the Poincaré group. The dimensions of $SO(2, 3)$ and the Poincaré group are ten. In general, the dimension of the isometry group, i.e., the number of independent Killing vectors, in D-dimensional spacetime is at most $\frac{1}{2}D(D + 1)$ (see, e.g., Chap.13 of [15]). Spacetime having an isometry of this dimension is called maximally symmetric. Minkowski spacetime and AdS spacetime are maximally symmetric. Another maximally symmetric spacetime is de Sitter spacetime, which is a solution of the Einstein equation with a positive cosmological constant and has the isometry $SO(1, 4)$. It is a general property of maximally symmetric spacetimes that the Riemann tensor can be expressed in terms of the metric as in (2.45). The coefficient on the right-hand side of (2.45) is zero for Minkowski spacetime, positive for de Sitter spacetime and negative for AdS spacetime.

The AdS spacetime solution has a global supersymmetry. For a solution of the field equations to preserve supersymmetry, the supertransformation (2.43) must vanish for that solution. When $\psi_\mu = 0$, the supertransformation of $e_\mu{}^a$ automatically vanishes. Requiring $\delta_Q \psi_\mu = 0$ we obtain the condition

$$\left(D_\mu + \frac{1}{2}m\gamma_\mu\right)\varepsilon = 0. \tag{2.46}$$

This is a partial differential equation on the transformation parameter ε. Spinors ε satisfying (2.46) are called Killing spinors. If Killing spinors exist, the solution is supersymmetric. The consistency of (2.46) requires the integrability condition

$$
\begin{aligned}
0 &= \left[D_\mu + \frac{1}{2}m\gamma_\mu, \ D_\nu + \frac{1}{2}m\gamma_\nu\right]\varepsilon \\
&= \frac{1}{4}\left(R_{\mu\nu}{}^{ab} + 2m^2 e_\mu{}^a e_\nu{}^b\right)\gamma_{ab}\varepsilon.
\end{aligned}
\tag{2.47}
$$

Using (2.45) we see that this condition is satisfied by AdS spacetime. Solutions of (2.46) indeed exist and were explicitly constructed in [1]. Hence, the AdS solution has a global supersymmetry. The supertransformation with Killing spinors as transformation parameters and the $SO(2, 3)$ transformation together form a closed superalgebra $OSp(1|4)$. This algebra is an analog of the super Poincaré algebra for Minkowski spacetime and is called a super anti de Sitter (AdS) algebra. We will discuss more general super AdS algebras in Sect. 3.5.

2.6 Extended Supersymmetries

So far we have considered $\mathcal{N} = 1$ supersymmetry, which contains a single Majorana spinor supercharge. More generally, we can consider $\mathcal{N}$-extended supersymmetry [11], which contains $\mathcal{N}$ Majorana spinor supercharges Q^i ($i = 1, 2, \ldots, \mathcal{N}$). The anticommutation relation of the supercharges of the $\mathcal{N}$-extended super Poincaré algebra is

$$\{Q_\alpha^i, \bar{Q}^{j\beta}\} = -2i\,(\gamma^\mu)_\alpha{}^\beta P_\mu \delta^{ij} + \delta_\alpha^\beta U^{ij} + i(\gamma_5)_\alpha{}^\beta V^{ij}, \tag{2.48}$$

where $U^{ij} = -U^{ji}$, $V^{ij} = -V^{ji}$ are generators called the central charges and commute with all the generators of the algebra. Other commutation relations have the same form as (2.3).

As in Sect. 2.1, we can find possible supermultiplets of this superalgebra. Let us consider massless supermultiplets with vanishing central charges. In this case we can

Table 2.1 Supergravity multiplets

h	$\mathcal{N}=1$	$\mathcal{N}=2$	$\mathcal{N}=3$	$\mathcal{N}=4$	$\mathcal{N}=5$	$\mathcal{N}=6$	$\mathcal{N}=7$	$\mathcal{N}=8$
$+2$	1	1	1	1	1	1	1	1
$+\frac{3}{2}$	1	2	3	4	5	6	$7+1$	8
$+1$		1	3	6	10	$15+1$	$21+7$	28
$+\frac{1}{2}$			1	4	$10+1$	$20+6$	$35+21$	56
0				$1+1$	$5+5$	$15+15$	$35+35$	70
$-\frac{1}{2}$			1	4	$1+10$	$6+20$	$21+35$	56
-1		1	3	6	10	$1+15$	$7+21$	28
$-\frac{3}{2}$	1	2	3	4	5	6	$1+7$	8
-2	1	1	1	1	1	1	1	1

construct $\mathcal{N}$ pairs of creation and annihilation operators $(b^i, b^{i\dagger})$ $(i = 1, 2, \ldots, \mathcal{N})$ satisfying the anticommutation relations of fermions. Therefore, a supermultiplet contains the states

$$|h_0\rangle, \quad b^{i\dagger}|h_0\rangle, \quad b^{i\dagger}b^{j\dagger}|h_0\rangle, \ldots, \quad b^{1\dagger}b^{2\dagger}\cdots b^{\mathcal{N}\dagger}|h_0\rangle, \qquad (2.49)$$

where $|h_0\rangle$ is a state with helicity h_0 and satisfying $b^i|h_0\rangle = 0$. Since $b^{i\dagger}$ has helicity $\frac{1}{2}$, helicities of these states are

$$h = h_0, \ h_0 + \tfrac{1}{2}, \ h_0 + 1, \ldots, \ h_0 + \tfrac{1}{2}\mathcal{N}. \qquad (2.50)$$

We see that for $\mathcal{N} > 8$ all the supermultiplets contain states with helicity $|h| > 2$. However, consistent interacting field theories are not known when they contain massless fields with helicity $|h| > 2$. As a consequence, $\mathcal{N} = 8$ is the largest supersymmetry that has been realized by field theories.

Supermultiplets which contain a graviton ($h = \pm 2$) and gravitinos ($h = \pm\frac{3}{2}$) are called supergravity multiplets. The numbers of states in supergravity multiplets are listed in Table 2.1. Since the states with helicity $h_0 + \frac{1}{2}n$ in (2.49) contain n anticommuting $b^{i\dagger}$, the number of such states is $_{\mathcal{N}}C_n$. In Table 2.1 we have added helicity flipped states required by the CPT theorem as in (2.10). Note that the $\mathcal{N} = 8$ supergravity multiplet contains all the states required by the CPT theorem without adding helicity flipped states. Note also that the $\mathcal{N} = 7$ and $\mathcal{N} = 8$ supergravity multiplets contain the same states and are expected to give the same theory.

One can construct extended supergravities corresponding to the supergravity multiplets in Table 2.1. The field contents of such theories are summarized in Table 3.3. Extended supergravities generically contain vector fields, spinor fields and scalar fields in addition to the gravitational field and the Rarita–Schwinger fields. As we mentioned above, field theories with $\mathcal{N} > 8$ supersymmetry are not known, and $\mathcal{N} = 8$ supergravity is called the maximal supergravity.

2.7 $\mathcal{N} = 2$ Poincaré Supergravity

The simplest extended supergravity is $\mathcal{N} = 2$ supergravity [4]. The field content is a gravitational field $e_\mu{}^a(x)$, a vector field $B_\mu(x)$ and two Majorana Rarita–Schwinger fields $\psi_\mu^i(x)$ $(i = 1, 2)$. The Lagrangian is

$$\mathcal{L} = e\hat{R} - \frac{1}{2}e\bar{\psi}_\mu^i \gamma^{\mu\nu\rho} \hat{D}_\nu \psi_\rho^i - \frac{1}{4}e F_{\mu\nu} F^{\mu\nu}$$
$$+ \frac{1}{16}e\varepsilon^{ij} \bar{\psi}_\mu^i \gamma^{[\mu} \gamma_{\rho\sigma} \gamma^{\nu]} \psi_\nu^j \left(F^{\rho\sigma} + \hat{F}^{\rho\sigma}\right), \tag{2.51}$$

where $F_{\mu\nu} = \partial_\mu B_\nu - \partial_\nu B_\mu$ is the field strength of the $U(1)$ gauge field B_μ and

$$\hat{F}_{\mu\nu} = F_{\mu\nu} - \frac{1}{2}\varepsilon^{ij} \bar{\psi}_\mu^i \psi_\nu^j \tag{2.52}$$

is the supercovariant field strength. ε^{ij} is the antisymmetric symbol with a component $\varepsilon^{12} = +1$. The scalar curvature $\hat{R}$ and the covariant derivative $\hat{D}_\mu$ are defined by using the spin connection $\hat{\omega}_{\mu ab}$ given by (2.23) with the replacements $\bar{\psi}_a \gamma_\mu \psi_b \to \bar{\psi}_a^i \gamma_\mu \psi_b^i$, etc. The covariant derivative does not contain a minimal coupling to the $U(1)$ gauge field B_μ. This means that the Rarita–Schwinger fields do not have a non-zero $U(1)$ charge. The coupling of the vector field and the Rarita–Schwinger fields is given by the last term of (2.51), which is called the Pauli term.

The Lagrangian (2.51) is invariant up to total divergences under the general coordinate transformation, the local Lorentz transformation and the $U(1)$ gauge transformation

$$\delta_g e_\mu{}^a = 0, \quad \delta_g B_\mu = \partial_\mu \zeta, \quad \delta_g \psi_\mu^i = 0. \tag{2.53}$$

It is also invariant under the $\mathcal{N} = 2$ local supertransformation

$$\delta_Q e_\mu{}^a = \frac{1}{4}\bar{\varepsilon}^i \gamma^a \psi_\mu^i, \qquad \delta_Q B_\mu = \frac{1}{2}\varepsilon^{ij} \bar{\varepsilon}^i \psi_\mu^j,$$
$$\delta_Q \psi_\mu^i = \hat{D}_\mu \varepsilon^i - \frac{1}{8}\varepsilon^{ij} \gamma^{\rho\sigma} \gamma_\mu \varepsilon^j \hat{F}_{\rho\sigma}, \tag{2.54}$$

where the transformation parameters are two Majorana spinors $\varepsilon^i(x)$ $(i = 1, 2)$. The commutation relation of two local supertransformations is

$$[\delta_Q(\varepsilon_1), \delta_Q(\varepsilon_2)] = \delta_G(\xi) + \delta_L(\lambda) + \delta_g(\zeta) + \delta_Q(\varepsilon), \tag{2.55}$$

where the transformation parameters on the right-hand side are

$$\xi^\mu = \frac{1}{4}\bar\varepsilon_2^i\gamma^\mu\varepsilon_1^i\,, \quad \lambda_{ab} = -\xi^\mu\hat\omega_{\mu ab} + \frac{1}{16}\varepsilon^{ij}\bar\varepsilon_2^i\gamma_{[a}\gamma^{\mu\nu}\gamma_{b]}\varepsilon_1^j\,\hat F_{\mu\nu},$$

$$\zeta = \frac{1}{2}\varepsilon^{ij}\bar\varepsilon_2^i\varepsilon_1^j - \xi^\mu B_\mu\,, \quad \varepsilon^i = -\xi^\mu\psi_\mu^i. \tag{2.56}$$

To obtain (2.55) we have to use the Rarita–Schwinger field equation.

This theory is called $\mathcal{N} = 2$ Poincaré supergravity since its field equations have a Minkowski spacetime solution $e_\mu{}^a = \delta_\mu^a$, $B_\mu = 0$, $\psi_\mu^i = 0$. For this background the commutator algebra of the local symmetry transformations reduces to the $\mathcal{N} = 2$ super Poincaré algebra. The $U(1)$ transformation $\delta_g(\zeta)$ on the right-hand side of (2.55) corresponds to the central charge in (2.48).

Global $U(2)$ Symmetry

This theory has a global $U(2)$ symmetry in addition to the above local symmetries [3]. This symmetry is an analog of the global $U(1)$ symmetry (2.31) of the $\mathcal{N} = 1$ theory. $SU(2)$ in $U(2) \sim SU(2) \times U(1)$ is a symmetry of the Lagrangian but the remaining $U(1)$ is a symmetry of the field equations.

The Lagrangian (2.51) is invariant under the global transformations with real constant parameters Σ^{ij}, Λ^{ij} satisfying $\Sigma^{ij} = -\Sigma^{ji}$, $\Lambda^{ij} = \Lambda^{ji}$, $\Lambda^{ii} = 0$:

$$\delta e_\mu{}^a = 0, \quad \delta B_\mu = 0, \quad \delta\psi_\mu^i = \left(\Sigma^{ij} + i\Lambda^{ij}\gamma_5\right)\psi_\mu^j. \tag{2.57}$$

These transformations form the group $SU(2)$. To see this, we decompose ψ_μ into the chirality eigenstates $\psi_{\mu\pm}^i = \frac{1}{2}(1\pm\gamma_5)\psi_\mu^i$. The transformation of the positive chirality component is $\delta\psi_{\mu+}^i = \left(\Sigma^{ij} + i\Lambda^{ij}\right)\psi_{\mu+}^j$. The 2×2 matrix $\Sigma + i\Lambda$ is a traceless anti-hermitian matrix and represents an infinitesimal $SU(2)$ transformation. The negative chirality component $\psi_{\mu-}^i$ is the charge conjugation of $\psi_{\mu+}^i$.

This theory also has a global $U(1)$ symmetry. This $U(1)$ is not a symmetry of the Lagrangian or the action but a symmetry of the field equations. The field equations of the vector field can be written as

$$\partial_\mu(e*G^{\mu\nu}) = 0, \quad \partial_\mu(e*F^{\mu\nu}) = 0, \tag{2.58}$$

where $*$ in $*F^{\mu\nu}$ and $*G^{\mu\nu}$ is the Hodge dual of second rank antisymmetric tensors defined as

$$*F^{\mu\nu} = \frac{1}{2}e^{-1}\varepsilon^{\mu\nu\rho\sigma}F_{\rho\sigma}. \tag{2.59}$$

$*G^{\mu\nu}$ in the first equation of (2.58) is defined by

$$*G^{\mu\nu} = \frac{2}{e}\frac{\partial\mathcal{L}}{\partial F_{\mu\nu}} = -F^{\mu\nu} + \frac{1}{4}\varepsilon^{ij}\bar\psi^{\rho i}\gamma_{[\rho}\gamma^{\mu\nu}\gamma_{\sigma]}\psi^{\sigma j}. \tag{2.60}$$

The first equation of (2.58) is the Euler equation derived from the Lagrangian, and the second one is the Bianchi identity, which implies that $F_{\mu\nu}$ can be expressed by the potential B_μ. These two equations correspond to Maxwell's equations of electromagnetism. Since the two equations in (2.58) have the same form, they are invariant under general linear transformations of $(F_{\mu\nu}, G_{\mu\nu})$. However, we should note that $F_{\mu\nu}$ and $G_{\mu\nu}$ are not independent, but are related by (2.60). Taking account of this relation the symmetry of (2.58) is the invariance under the $U(1)$ transformation with a real constant parameter Λ:

$$\delta e_\mu{}^a = 0, \quad \delta \begin{pmatrix} F_{\mu\nu} \\ G_{\mu\nu} \end{pmatrix} = \begin{pmatrix} 0 & \Lambda \\ -\Lambda & 0 \end{pmatrix} \begin{pmatrix} F_{\mu\nu} \\ G_{\mu\nu} \end{pmatrix}, \quad \delta\psi_\mu^i = \frac{1}{2} i\Lambda\gamma_5\psi_\mu^i. \tag{2.61}$$

The definition of $G_{\mu\nu}$ in (2.60) and the field equations of $e_\mu{}^a$, ψ_μ^i are also invariant under (2.61). The second equation in (2.61) represents an interchange of $F_{\mu\nu}$ and $G_{\mu\nu} = *F_{\mu\nu} + \cdots$, i.e., an interchange of the electric field and the magnetic field. In general, a symmetry which exchanges the field equation and the Bianchi identity as in (2.61) is called the duality symmetry. The duality symmetry also appears in other even-dimensional supergravities and plays a crucial role in applications to string theory. We will discuss the duality symmetry in detail in Sect. 4.2.

Relation to the $\mathcal{N} = 1$ Theory

The Lagrangian of $\mathcal{N} = 1$ supergravity (2.21) can be obtained from that of $\mathcal{N} = 2$ supergravity (2.51) by imposing the conditions $B_\mu = 0$, $\psi_\mu^2 = 0$. Under the supertransformation with the parameter ε^1 in (2.54) these conditions are preserved and the remaining fields $e_\mu{}^a$, ψ_μ^1 transform as in the $\mathcal{N} = 1$ transformation (2.28). Thus, we have obtained the $\mathcal{N} = 1$ theory from the $\mathcal{N} = 2$ theory.

In general, we can obtain the $\mathcal{N}'$-extended theory from the $\mathcal{N}$-extended theory $(\mathcal{N} > \mathcal{N}')$ by a truncation, i.e., by putting a certain set of the fields equal to zero. The truncation must be consistent with the field equations and the supertransformation. The field equations of the fields which we put to zero must be automatically satisfied. Furthermore, the $\mathcal{N}'$-extended supertransformation of the fields which we put to zero must automatically vanish.

2.8 $\mathcal{N} = 2$ Anti de Sitter Supergravity

We can introduce a cosmological term to $\mathcal{N} = 2$ supergravity [6, 7]. The Lagrangian is

$$\mathcal{L} = e\hat{R} + 6m^2 e - \frac{1}{2} e\bar{\psi}_\mu^i \gamma^{\mu\nu\rho} \hat{D}_\nu \psi_\rho^i + \frac{1}{2} me\bar{\psi}_\mu^i \gamma^{\mu\nu} \psi_\nu^i$$
$$- \frac{1}{4} e F_{\mu\nu} F^{\mu\nu} + \frac{1}{16} e\varepsilon^{ij} \bar{\psi}_\mu^i \gamma^{[\mu} \gamma_{\rho\sigma} \gamma^{\nu]} \psi_\nu^j \left(F^{\rho\sigma} + \hat{F}^{\rho\sigma} \right). \tag{2.62}$$

As in the $\mathcal{N} = 1$ case the cosmological term with a negative cosmological constant and the mass term of the Rarita–Schwinger fields are added. In addition to these modifications the covariant derivative on the Rarita–Schwinger fields

$$D_{[\mu} \psi^i_{\nu]} = \left(\partial_{[\mu} + \frac{1}{4} \hat{\omega}_{[\mu}{}^{ab} \gamma_{ab} \right) \psi^i_{\nu]} + \frac{1}{2} m \varepsilon^{ij} B_{[\mu} \psi^j_{\nu]} \qquad (2.63)$$

contains a minimal coupling to the $U(1)$ gauge field B_μ, which is not present in the $m = 0$ theory. This corresponds to a gauging of an $SO(2)$ subgroup of the global symmetry $U(2)$ (the Σ^{ij} transformation in (2.57)) of the $m = 0$ theory. The gauge coupling constant is $g = \frac{1}{2} m$ and is proportional to the parameter m appearing in the cosmological term and the Rarita–Schwinger mass term.

This Lagrangian is invariant up to total divergences under the general coordinate transformation, the local Lorentz transformation and the $U(1)$ gauge transformation

$$\delta_g e_\mu{}^a = 0, \quad \delta_g B_\mu = \partial_\mu \zeta, \quad \delta_g \psi^i_\mu = -\frac{1}{2} m \zeta \varepsilon^{ij} \psi^j_\mu. \qquad (2.64)$$

It is also invariant under the $\mathcal{N} = 2$ local supertransformation

$$\delta_Q e_\mu{}^a = \frac{1}{4} \bar{\varepsilon}^i \gamma^a \psi^i_\mu, \qquad \delta_Q B_\mu = \frac{1}{2} \varepsilon^{ij} \bar{\varepsilon}^i \psi^j_\mu,$$

$$\delta_Q \psi^i = \hat{D}_\mu \varepsilon^i + \frac{1}{2} m \gamma_\mu \varepsilon^i - \frac{1}{8} \varepsilon^{ij} \gamma^{\rho\sigma} \gamma_\mu \varepsilon^j \hat{F}_{\rho\sigma}. \qquad (2.65)$$

The covariant derivative on ε^i contains a minimal coupling to B_μ as in (2.63). The commutation relation of two local supertransformations has the same form as (2.55). The parameters on the right-hand side are the same as (2.56) except that the local Lorentz transformation parameter has the additional term $-\frac{1}{4} m \bar{\varepsilon}_2{}^i \gamma_{ab} \varepsilon^i_1$. The global $U(2)$ symmetry of the $m = 0$ theory is broken by the coupling to the gauge field since only $U(1)$ part of $U(2)$ was gauged.

This theory is called $\mathcal{N} = 2$ anti de Sitter (AdS) supergravity since its field equations have an AdS spacetime solution. It is also called $\mathcal{N} = 2$ gauged supergravity since it has the minimal coupling to the gauge field.

2.9 $\mathcal{N} \geq 3$ Supergravities

Similarly, $\mathcal{N} = 3$, 4, 5, 6, 8 extended supergravities can be constructed. Let us have a quick look at these theories.

$\mathcal{N} = 3$ Poincaré supergravity is similar to the $\mathcal{N} = 2$ theory. The field content is a gravitational field, three Majorana Rarita–Schwinger fields, three vector fields and a Majorana spinor field. The action is invariant under the general coordinate transformation, the local Lorentz transformation, the $U(1)^3$ gauge transformation and the $\mathcal{N} = 3$ local supertransformation. The field equations are invariant under the global $U(3)$ transformation, which includes the duality transformation of the vector fields.

One can also construct $\mathcal{N} = 3$ gauged (AdS) supergravity. The Lagrangian has a cosmological term and a mass term of the Rarita–Schwinger fields as in (2.42) of the $\mathcal{N} = 1$ theory. A new feature of the $\mathcal{N} = 3$ theory is that the three vector fields become the non-Abelian $SO(3)$ Yang–Mills field. This corresponds to a gauging of a subgroup $SO(3)$ of the global $U(3)$ in the ungauged theory. The Rarita–Schwinger fields have a minimal coupling to the Yang–Mills field. The gauge coupling constant is $g = \frac{1}{2}m$ as in the $\mathcal{N} = 2$ theory.

$\mathcal{N} \geq 4$ Poincaré supergravities contain scalar fields and are much different from the $\mathcal{N} < 4$ theories. The field content is a gravitational field, $\mathcal{N}$ Majorana Rarita–Schwinger fields, $\frac{1}{2}\mathcal{N}(\mathcal{N}-1)$ vector fields and a certain number of scalar and spinor fields. The scalar fields have non-polynomial interactions. The action is invariant under the general coordinate transformation, the local Lorentz transformation, the $U(1)^{\frac{1}{2}\mathcal{N}(\mathcal{N}-1)}$ gauge transformation and the $\mathcal{N}$ local supertransformation.

A significant feature of the $\mathcal{N} \geq 4$ theories is a global symmetry of a non-compact Lie group G. For instance, the $\mathcal{N} = 4$ theory has a global $G = SU(4) \times SU(1, 1)$ symmetry. This symmetry is an analog of $U(\mathcal{N})$ symmetry in the $\mathcal{N} \leq 3$ theories. The $U(1)$ subgroup of $U(4) = SU(4) \times U(1)$ is enlarged to the non-compact group $SU(1, 1)$. This G symmetry is a symmetry of the field equations since it contains the duality transformation. In ordinary field theories like the standard theory of particle physics the internal symmetry is a compact Lie group. However, $\mathcal{N} \geq 4$ supergravities can have a non-compact internal symmetry since they contain scalar fields represented by a non-linear sigma model. The non-linear sigma model is a theory of scalar fields which take values in a coset space G/H, where G is a Lie group and H is a subgroup of G. Originally, non-linear sigma models were introduced in order to describe massless Nambu–Goldstone bosons when a symmetry G is spontaneously broken to H. In supergravities G is a non-compact group and H is a maximal compact subgroup of G. We will discuss non-linear sigma models and non-compact symmetries in Chap. 4.

One can also construct gauged supergravities for $\mathcal{N} \geq 4$, which contain minimal couplings to the vector fields. The $\frac{1}{2}\mathcal{N}(\mathcal{N} - 1)$ vector fields in the theory become the Yang–Mills field with the gauge group $SO(\mathcal{N})$, which is a subgroup of the non-compact symmetry G of the ungauged theory. (It is also possible to gauge other (non-compact) subgroup of G.) The Lagrangian contains a potential term of the scalar fields $-g^2 eV(\phi)$ and Yukawa couplings of the fermionic fields and the scalar fields such as $gef(\phi)\bar{\psi}_\mu\gamma^{\mu\nu}\psi_\nu$. Here, g is the gauge coupling constant of the Yang–Mills field. These terms effectively become a cosmological term and fermion mass terms when the scalar fields have vacuum expectation values for which $V(\phi)$ and $f(\phi)$ are non-zero. Then, the field equations have an AdS spacetime solution. We will discuss gauged supergravities in Chap. 7.

References

1. P. Breitenlohner, D.Z. Freedman, Stability in gauged extended supergravity. Ann. Phys. **144**, 249 (1982)
2. S. Deser, B. Zumino, Consistent supergravity. Phys. Lett. **B62**, 335 (1976)
3. S. Ferrara, J. Scherk, B. Zumino, Algebraic properties of extended supergravity theories. Nucl. Phys. **B121**, 393 (1977)
4. S. Ferrara, P. van Nieuwenhuizen, Consistent supergravity with complex spin-$\frac{3}{2}$ gauge fields. Phys. Rev. Lett. **37**, 1669 (1976)
5. S. Ferrara, P. van Nieuwenhuizen, The auxiliary fields of supergravity. Phys. Lett. **B74**, 333 (1978)
6. E.S. Fradkin and M.A. Vasiliev, Model of supergravity with minimal electromagnetic interaction, Lebedev Institute preprint LEBEDEV-76-197 (1976)
7. D.Z. Freedman, A. Das, Gauge internal symmetry in extended supergravity. Nucl. Phys. **B120**, 221 (1977)
8. D.Z. Freedman, A. van Proeyen, *Supergravity* (Cambridge University Press, Cambridge, 2012)
9. D.Z. Freedman, P. van Nieuwenhuizen, Properties of supergravity theory. Phys. Rev. **D14**, 912 (1976)
10. D.Z. Freedman, P. van Nieuwenhuizen, S. Ferrara, Progress toward a theory of supergravity. Phys. Rev. **D13**, 3214 (1976)
11. R. Haag, J.T. Łopuszański, M. Sohnius, All possible generators of supersymmetries of the S matrix. Nucl. Phys. **B88**, 257 (1975)
12. M.F. Sohnius, P.C. West, An alternative minimal off-shell version of $\mathcal{N} = 1$ supergravity. Phys. Lett. **B105**, 353 (1981)
13. K.S. Stelle, P.C. West, Minimal auxiliary fields for supergravity. Phys. Lett. **B74**, 330 (1978)
14. P.K. Townsend, Cosmological constant in supergravity. Phys. Rev. **D15**, 2802 (1977)
15. S. Weinberg, *Gravitation and Cosmology* (Wiley, New York, 1972)
16. J. Wess, J. Bagger, *Supersymmetry and Supergravity*, 2nd edn. (Princeton University Press, Princeton, 1992)

Chapter 3
Superalgebras and Supermultiplets

3.1 Spinors in General Dimensions

To construct supergravities in higher dimensions we need to know what types
of spinors can be defined in each dimension [4, 10]. Spinors in D dimensions
are quantities which transform in a spinor representation of the Lorentz group
$SO(1, D - 1)$. Here, we consider spinor representations of more general group
$SO(t, s)$ with the invariant metric

$$\eta_{ab} = \text{diag}\,(\underbrace{-1, \ldots, -1}_{t}, \underbrace{+1, \ldots, +1}_{s}) \qquad (t + s = D)\,. \tag{3.1}$$

The D-dimensional Lorentz group $SO(1, D - 1)$ and the D-dimensional rotation
group $SO(0, D) = SO(D)$ are special cases of $SO(t, s)$.

3.1.1 Gamma Matrices

The gamma matrices γ^a $(a = 1, 2, \ldots, D)$[1] of $SO(t, s)$ satisfy the anticommutation
relation

$$\{\gamma^a, \gamma^b\} = 2\eta^{ab}\,\mathbf{1}\,, \tag{3.2}$$

where $\mathbf{1}$ is the unit matrix. The matrices γ^a are anti-hermitian for $a = 1, \ldots, t$ and
hermitian for $a = t + 1, \ldots, D$. The smallest matrices satisfying this anticommutation relation are $2^{[D/2]} \times 2^{[D/2]}$, where $[x]$ is the largest integer not larger than x. An
explicit representation of the gamma matrices can be constructed as tensor products
of 2×2 matrices. In the case of $SO(2n)$ we can use the representation

---

[1] For the Lorentz group $SO(1, D - 1)$ we use $a = 0, 1, 2, \ldots, D - 1$.

Y. Tanii, *Introduction to Supergravity*, SpringerBriefs in Mathematical Physics,
DOI: 10.1007/978-4-431-54828-7_3, © The Author(s) 2014

$$\gamma^1 = \sigma_1 \otimes \mathbf{1} \otimes \cdots \otimes \mathbf{1},$$
$$\gamma^2 = \sigma_2 \otimes \mathbf{1} \otimes \cdots \otimes \mathbf{1},$$
$$\gamma^3 = \sigma_3 \otimes \sigma_1 \otimes \mathbf{1} \otimes \cdots \otimes \mathbf{1},$$
$$\gamma^4 = \sigma_3 \otimes \sigma_2 \otimes \mathbf{1} \otimes \cdots \otimes \mathbf{1},$$

$$\vdots$$

$$\gamma^{2k+1} = \underbrace{\sigma_3 \otimes \cdots \otimes \sigma_3}_{k} \otimes \sigma_1 \otimes \mathbf{1} \otimes \cdots \otimes \mathbf{1},$$

$$\gamma^{2k+2} = \underbrace{\sigma_3 \otimes \cdots \otimes \sigma_3}_{k} \otimes \sigma_2 \otimes \mathbf{1} \otimes \cdots \otimes \mathbf{1},$$

$$\vdots$$

$$\gamma^{2n-1} = \sigma_3 \otimes \cdots \otimes \sigma_3 \otimes \sigma_1,$$
$$\gamma^{2n} = \sigma_3 \otimes \cdots \otimes \sigma_3 \otimes \sigma_2, \tag{3.3}$$

where $\mathbf{1}$ is the 2×2 unit matrix and σ_i $(i = 1, 2, 3)$ are the Pauli matrices (2.5). The gamma matrices of $SO(t, s)$ $(t + s = 2n)$ can be obtained from those of $SO(2n)$ by multiplying the first t matrices by the imaginary unit i. The gamma matrices of $SO(t, s)$ $(t + s = 2n + 1)$ can be obtained from those of $SO(t, s - 1)$ as follows. Use the gamma matrices γ^a $(a = 1, \ldots, 2n)$ of $SO(t, s - 1)$ as the first $2n$ matrices of $SO(t, s)$. As the last matrix use the matrix

$$\gamma^{2n+1} = (-1)^{\frac{1}{4}(s-t-1)} \gamma^1 \gamma^2 \cdots \gamma^{2n}. \tag{3.4}$$

One can easily see that these $2n + 1$ matrices satisfy the anticommutation relation of $SO(t, s)$ in (3.2). Various formulae of the gamma matrices are given in Appendix B.

The following Pauli's fundamental theorem is useful below. When $D = $ even, if two sets of D $2^{D/2} \times 2^{D/2}$ matrices $\{\gamma^a\}$ and $\{\gamma'^a\}$ satisfy the anticommutation relation (3.2), i.e., $\{\gamma^a, \gamma^b\} = 2\eta^{ab}\mathbf{1}$, $\{\gamma'^a, \gamma'^b\} = 2\eta^{ab}\mathbf{1}$, then there exists a matrix S with the property

$$\gamma'^a = S\gamma^a S^{-1}. \tag{3.5}$$

Such an S is unique up to a constant multiple. A proof of this theorem for $D = 4$ can be found in Appendix C of [9], which can be easily extended to general even D.

3.1.2 Dirac Spinors

General spinors of $SO(t, s)$ have $2^{[D/2]}$ complex components and transform under the infinitesimal $SO(t, s)$ transformation with a parameter $\lambda_{ab} = -\lambda_{ba}$ as

$$\delta\psi = -\frac{1}{4}\lambda_{ab}\gamma^{ab}\psi \, , \tag{3.6}$$

where $\gamma^{ab} = \gamma^{[a}\gamma^{b]}$ is an antisymmetrized product of two gamma matrices. We define the matrix A as

$$A = (-1)^{\frac{1}{4}t(t+1)}\gamma^1\gamma^2\ldots\gamma^t \, , \tag{3.7}$$

which satisfies

$$(\gamma^a)^\dagger = (-1)^t A\gamma^a A^{-1} \, , \quad A^\dagger = A \, , \quad A^2 = \mathbf{1} \, . \tag{3.8}$$

By using this matrix we define the Dirac conjugate of a spinor ψ as

$$\bar{\psi} = \psi^\dagger A \, . \tag{3.9}$$

For $SO(1, D-1)$ it becomes $\bar{\psi} = \psi^\dagger i\gamma^0$, which is the usual definition of the Dirac conjugate. The Dirac conjugate (3.9) transforms under $SO(t, s)$ as

$$\delta\bar{\psi} = \frac{1}{4}\lambda_{ab}\bar{\psi}\gamma^{ab} \, . \tag{3.10}$$

We can construct bilinear forms $\bar{\psi}\psi$, $\bar{\psi}\gamma^a\psi$, $\bar{\psi}\gamma^{ab}\psi$, $\ldots$, which transform as a scalar, a vector, a second rank antisymmetric tensor, $\ldots$ under $SO(t, s)$.

As we will discuss below, we can reduce the number of independent components of spinors by imposing certain conditions. Such conditions must be consistent with the $SO(t, s)$ transformation (3.6). General spinors without any condition imposed are called Dirac spinors. To discuss supersymmetry it is convenient to use spinors with the smallest number of independent components in each dimension.

3.1.3 Weyl Spinors

Weyl spinors are spinors having a definite chirality and can be defined only when D is even. We define the chirality of spinors as an eigenvalue of the chirality matrix

$$\bar{\gamma} = (-1)^{\frac{1}{4}(s-t)}\gamma_1\gamma_2\ldots\gamma_D \, , \tag{3.11}$$

which satisfies

$$\bar{\gamma}^2 = 1, \quad \{\bar{\gamma}, \gamma^a\} = 0, \quad \bar{\gamma}^\dagger = \bar{\gamma} \tag{3.12}$$

and has eigenvalues ± 1. This matrix $\bar{\gamma}$ is a generalization of $\gamma_5 = i\gamma_0\gamma_1\gamma_2\gamma_3$ for the Lorentz group $SO(1, 3)$. Weyl spinors with positive chirality ψ_+ and those with negative chirality ψ_- are defined by the Weyl conditions

$$\bar{\gamma}\psi_+ = \psi_+, \quad \bar{\gamma}\psi_- = -\psi_- . \tag{3.13}$$

It is easy to see that these conditions are consistent with (3.6) since γ^{ab} commutes with $\bar{\gamma}$. The Dirac conjugates satisfy

$$\bar{\psi}_+\bar{\gamma} = (-1)^t \bar{\psi}_+, \quad \bar{\psi}_-\bar{\gamma} = -(-1)^t \bar{\psi}_- . \tag{3.14}$$

The chirality matrix $\bar{\gamma}$ and therefore Weyl spinors can be defined for any even D. For odd D, $\bar{\gamma}$ is proportional to the unit matrix since

$$\gamma_1\gamma_2 \cdots \gamma_D = (-1)^{-\frac{1}{4}(s+3t-1)} \mathbf{1} \tag{3.15}$$

as can be seen from (3.4), and Weyl spinors cannot be defined.

3.1.4 Majorana Spinors

Majorana spinors are spinors satisfying the Majorana condition

$$\psi^c = \psi, \tag{3.16}$$

where ψ^c is the charge conjugation of ψ. This is a kind of reality condition. We shall discuss the definitions of the charge conjugation for $D = $ even and for $D = $ odd separately.

Let us first consider the case $D = t + s = $ even. Since the matrices $\pm(\gamma^a)^*$ satisfy the same anticommutation relation as γ^a, by Pauli's fundamental theorem there exist matrices B_+ and B_- satisfying

$$(\gamma^a)^* = B_+\gamma^a B_+^{-1} , \quad -(\gamma^a)^* = B_-\gamma^a B_-^{-1} . \tag{3.17}$$

The charge conjugation is defined by using one of these matrices as

$$\psi^c = B_+^{-1}\psi^* \quad \text{or} \quad \psi^c = B_-^{-1}\psi^* . \tag{3.18}$$

By (3.17), ψ^c transforms under $SO(t, s)$ in the same way as ψ in (3.6). It can be shown that $B_\pm$ satisfy

Table 3.1 Values of $\varepsilon_\pm$

$s - t$ mod 8	1	2	3	4	5	6	7	8
ε_+	+1	+1	No	−1	−1	−1	No	+1
ε_-	No	−1	−1	−1	No	+1	+1	+1

$$B_\pm^\dagger B_\pm = 1 , \qquad B_\pm^* B_\pm = \varepsilon_\pm 1 , \qquad B_\pm^T = \varepsilon_\pm B_\pm , \qquad (3.19)$$

where $\varepsilon_\pm$ are constants taking values ± 1 and are given by

$$\varepsilon_\pm = \sqrt{2} \cos\left[\frac{\pi}{4}(s - t \mp 1)\right] . \qquad (3.20)$$

(For proofs of (3.19), (3.20), see [4, 10].) We have summarized the values of $\varepsilon_\pm$ in Table 3.1.

The definitions of the charge conjugation (3.18) can be rewritten by using the charge conjugation matrix. The charge conjugation matrices $C_\pm$ are defined by using $B_\pm$ and A in (3.7) as

$$C_\pm = B_\pm^{-1} A^{-1T} . \qquad (3.21)$$

In terms of $C_\pm$, (3.18) can be rewritten as

$$\psi^c = C_+ \bar{\psi}^T \quad \text{or} \quad \psi^c = C_- \bar{\psi}^T , \qquad (3.22)$$

where $\bar{\psi}$ is the Dirac conjugate defined by (3.9). By (3.7), (3.8), (3.17), (3.19), the charge conjugation matrices $C_\pm$ satisfy

$$\gamma^{aT} = \pm(-1)^t C_\pm^{-1} \gamma^a C_\pm , \quad C_\pm^T = (\pm 1)^t (-1)^{\frac{1}{2}t(t+1)} \varepsilon_\pm C_\pm , \quad C_\pm^\dagger C_\pm = 1 . \qquad (3.23)$$

The charge conjugation matrix C of $SO(1, 3)$ used in (2.1), (2.2) is $C = C_+$.

For the Majorana condition (3.16) to be consistent we must have $(\psi^c)^c = \psi$, which is equivalent to $B_+^* B_+ = 1$ or $B_-^* B_- = 1$. Therefore, for $D =$ even, Majorana spinors can be defined only when $\varepsilon_+(t, s) = 1$ or $\varepsilon_-(t, s) = 1$. To distinguish the two cases, spinors satisfying (3.16) with the charge conjugation defined by using B_+ are called Majorana spinors, while those using B_- are called pseudo Majorana spinors. From Table 3.1 we can see which $SO(t, s)$ allows (pseudo) Majorana spinors.

The charge conjugation for $D = t + s =$ odd is defined by using the matrices $B_\pm$ for even $D - 1$. Recall that the gamma matrices of $SO(t, s)$ ($D = t + s =$ odd) can be constructed from those of $SO(t, s - 1)$. The first $D - 1$ matrices γ^a ($a = 1, 2, \ldots, D - 1$) are those of $SO(t, s - 1)$ and the last matrix γ^D is (3.4). Then, the matrices $B_\pm$ of $SO(t, s - 1)$ satisfy

Table 3.2 Possible types of spinors for $SO(1, D-1)$

D mod 8	D	W	M	pM	MW	pMW
11	o		o			
10	o	o	o	o	o	o
9	o			o		
8	o	o		o		
7	o					
6	o	o			•	
5	o					
4	o	o	o			

$$B_{\pm}\gamma^a B_{\pm}^{-1} = \pm(\gamma^a)^* \quad (a = 1, \ldots, D-1) \, ,$$

$$B_{\pm}\gamma^D B_{\pm}^{-1} = (-1)^{\frac{1}{2}(s-t-1)}(\gamma^D)^* \, . \tag{3.24}$$

When $(-1)^{\frac{1}{2}(s-t-1)} = +1$, we can use B_+ to define the charge conjugation since the signs on the right hand side in (3.24) are the same for all γ^a ($a = 1, \ldots, D$). Similarly, when $(-1)^{\frac{1}{2}(s-t-1)} = -1$, we can use B_- to define the charge conjugation. As in the case $D =$ even, the charge conjugation must satisfy $(\psi^c)^c = \psi$, i.e., $\varepsilon_+ = 1$ or $\varepsilon_- = 1$, in order to define (pseudo) Majorana spinors. Possible $B_{\pm}$ and corresponding $\varepsilon_{\pm}$ are listed in Table 3.1.

3.1.5 Majorana–Weyl Spinors

We can also define (pseudo) Majorana–Weyl spinors, which satisfy both of the (pseudo) Majorana condition $\psi^c = \psi$ and the Weyl condition $\bar{\gamma}\psi = \psi$ (or $\bar{\gamma}\psi = -\psi$). (Pseudo) Majorana–Weyl spinors are possible only if these two conditions are compatible, i.e., ψ and ψ^c have the same chirality. In general, when a spinor ψ has the chirality $\pm$, its charge conjugation ψ^c has the chirality $\pm(-1)^{\frac{1}{2}(s-t)}$. Thus, ψ and ψ^c have the same chirality only when $s - t = 0$ mod 4. As can be seen from $\varepsilon_{\pm}$ in Table 3.1 (pseudo) Majorana spinors cannot be defined when $s - t = 4$ mod 8. Therefore, (pseudo) Majorana–Weyl spinors can be defined only when $s - t = 8$ mod 8. In particular, they can be defined in $D = 2$ mod 8 for the Lorentz group $SO(1, D-1)$. Possible types of spinors in each dimension for $SO(1, D-1)$ are summarized in Table 3.2. In this Table D, W, M, pM, MW, pMW denote Dirac, Weyl, Majorana, pseudo Majorana, Majorana–Weyl, pseudo Majorana–Weyl spinors, respectively.

3.1.6 Symplectic Majorana Spinors

When $(\psi^c)^c = -\psi$, we cannot impose the (pseudo) Majorana condition $\psi^c = \psi$ and we have to use Dirac spinors (or Weyl spinors in even dimensions). Alternatively, we can introduce even numbers of spinors ψ^i ($i = 1, 2, \ldots, 2n$) and impose the symplectic (pseudo) Majorana condition

$$\psi^i = \Omega^{ij}(\psi^j)^c \,, \tag{3.25}$$

where Ω^{ij} is a constant antisymmetric unitary matrix satisfying $\Omega^{ij} = -\Omega^{ji}$, $\Omega^{ik}(\Omega^{jk})^* = \delta^i_j$. Because of the presence of Ω^{ij} in (3.25) this condition is consistent. Spinors satisfying this condition are called symplectic (pseudo) Majorana spinors. $2n$ symplectic (pseudo) Majorana spinors are equivalent to n Dirac spinors. Sometimes it is more convenient to use symplectic (pseudo) Majorana spinors than Dirac spinors, especially when the theory has a symplectic symmetry. When $s - t = 4$ mod 8, we can define symplectic (pseudo) Majorana–Weyl spinors satisfying both of the symplectic (pseudo) Majorana condition (3.25) and the Weyl condition (3.13).

3.2 Super Poincaré Algebras

Let us consider the super Poincaré algebras in general dimensions [11]. The super Poincaré algebra consists of the translation generators P_a, the Lorentz generators M_{ab}, the supercharges Q^i_α and the central charges Z^{ij}. The (anti)commutation relations of the generators are

$$[M_{ab}, M_{cd}] = -i\eta_{bc}M_{ad} + i\eta_{bd}M_{ac} + i\eta_{ac}M_{bd} - i\eta_{ad}M_{bc} \,,$$
$$[M_{ab}, P_c] = -i\eta_{bc}P_a + i\eta_{ac}P_b \,, \quad [P_a, P_b] = 0 \,,$$
$$[M_{ab}, Q^i] = \frac{1}{2}i\,\gamma_{ab}Q^i \,, \quad [P_a, Q^i] = 0 \tag{3.26}$$

and $\{Q, Q\}$ to be discussed below. The first three commutation relations are those of the Poincaré algebra. The central charges Z^{ij} commute with all the generators. There are transformations of these generators which preserve the above (anti)commutation relations. Such transformations are called the automorphisms and form a automorphism group K. The generators T_I ($I = 1, 2, \ldots, \dim K$) of K satisfy

$$[T_I, T_J] = i f_{IJ}{}^K T_K \,, \quad [T_I, Q^i] = -(t_I)^i{}_j Q^j \,,$$
$$[T_I, Z^{ij}] = -(t_I)^i{}_k Z^{kj} - (t_I)^j{}_k Z^{ik} \,, \tag{3.27}$$

where $f_{IJ}{}^K$ and t_I are the structure constant and representation matrices of the Lie algebra of K.

The automorphism group K and the anticommutator $\{Q, Q\}$ depend on the spinor type of Q^i.

- **$D = 4, \ 8 \bmod 8$**

The supercharges are Weyl spinors with negative chirality Q^i_- ($i = 1, 2, \ldots, \mathcal{N}$). (Alternatively, one can use (pseudo) Majorana spinors, since a (pseudo) Majorana spinor is equivalent to a Weyl spinor in these dimensions. However, it is more convenient to use Weyl spinors in order to write down anticommutation relations.) Their charge conjugations have positive chirality $(Q^i_-)^c = Q_{+i}$, where the charge conjugation matrix $C = C_+$ ($C = C_-$) is used for $D = 4$ ($D = 8$) mod 8. The anticommutation relations of the supercharges are

$$\{Q^i_-, Q^T_{+j}\} = -2\mathrm{i} \, \frac{1}{2}(1 - \bar{\gamma})\gamma^a C^T P_a \delta^i_j \ ,$$

$$\{Q^i_-, Q^{jT}_-\} = \frac{1}{2}(1 - \bar{\gamma})C^T Z^{ij} \ . \tag{3.28}$$

By (anti)symmetry of the charge conjugation matrix, the central charges satisfy $Z^{ij} = -Z^{ji}$ for $D = 4$ mod 8 and $Z^{ij} = Z^{ji}$ for $D = 8$ mod 8. The automorphism group is $K = U(\mathcal{N}) = SU(\mathcal{N}) \times U(1)$. Q^i_- and Q_{+i} belong to the fundamental representations $\mathcal{N}$ and $\bar{\mathcal{N}}$ of $SU(\mathcal{N})$, and have $U(1)$ charges $+1$ and -1, respectively. The anticommutation relation (2.48) for $D = 4$ can be rewritten in the form (3.28) with $Z^{ij} = U^{ij} - \mathrm{i}V^{ij}$ by using the Weyl spinor supercharges.

- **$D = 10 \bmod 8$**

The supercharges are Majorana–Weyl spinors Q^i_+ ($i = 1, 2, \ldots, \mathcal{N}_+$) with positive chirality and Q^i_- ($i = 1, 2, \ldots, \mathcal{N}_-$) with negative chirality. These are independent spinors, which are not related by the charge conjugation. One can use either of C_+ and C_- to define the charge conjugation. The anticommutation relations of the supercharges are

$$\{Q^i_\pm, Q^{jT}_\pm\} = -2\mathrm{i} \, \frac{1}{2}(1 \pm \bar{\gamma})\gamma^a C^T P_a \delta^{ij} \ ,$$

$$\{Q^i_+, Q^{jT}_-\} = \frac{1}{2}(1 + \bar{\gamma})C^T Z^{ij} \ . \tag{3.29}$$

The automorphism group is $K = SO(\mathcal{N}_+) \times SO(\mathcal{N}_-)$. Q^i_+ and Q^i_- belong to the representations $(\mathcal{N}_+, \mathbf{0})$ and $(\mathbf{0}, \mathcal{N}_-)$ of K, respectively. We call the supersymmetry with this algebra as $\mathcal{N} = (\mathcal{N}_+, \mathcal{N}_-)$.

- **$D = 6 \bmod 8$**

The supercharges are symplectic Majorana–Weyl spinors Q^i_+ ($i = 1, 2, \ldots, \mathcal{N}_+$) with positive chirality and Q^i_- ($i = 1, 2, \ldots, \mathcal{N}_-$) with negative chirality. They satisfy $\Omega^{ij}_+(Q^j_+)^c = Q^i_+$, $\Omega^{ij}_-(Q^j_-)^c = Q^i_-$, where $\Omega^{ij}_\pm$ are constant antisymmetric unitary matrices. The numbers $\mathcal{N}_+$ and $\mathcal{N}_-$ must be even. Q^i_+ and Q^i_- are independent spinors. One can use either of C_+ and C_- to define the charge conjugation. The anticommutation relations of the supercharges are

$$\{Q^i_\pm, Q^{jT}_\pm\} = -2\mathrm{i}\,\frac{1}{2}(1 \pm \bar{\gamma})\gamma^a C^T P_a \Omega^{ij}_\pm \,,$$

$$\{Q^i_+, Q^{jT}_-\} = \frac{1}{2}(1 + \bar{\gamma})C^T Z^{ij} \,. \tag{3.30}$$

The automorphism group is $K = USp(\mathcal{N}_+) \times USp(\mathcal{N}_-)$. Q^i_+ and Q^i_- belong to the representations $(\mathcal{N}_+, \mathbf{0})$ and $(\mathbf{0}, \mathcal{N}_-)$ of K, respectively. We call the supersymmetry with this algebra as $\mathcal{N} = (\mathcal{N}_+, \mathcal{N}_-)$.

- $D = 9,\ 11 \bmod 8$

The supercharges are (pseudo) Majorana spinors Q^i ($i = 1, 2, \ldots, \mathcal{N}$). The anticommutation relations of the supercharges are

$$\{Q^i, Q^{jT}\} = -2\mathrm{i}\gamma^a C^T P_a \delta^{ij} + C^T Z^{ij} \,, \tag{3.31}$$

where $C = C_-$, $Z^{ij} = Z^{ji}$ for $D = 9 \bmod 8$ and $C = C_+$, $Z^{ij} = -Z^{ji}$ for $D = 11 \bmod 8$. The automorphism group is $K = SO(\mathcal{N})$ and Q^i belong to the representation $\mathcal{N}$ of K.

- $D = 5,\ 7 \bmod 8$

The supercharges are symplectic (pseudo) Majorana spinors Q^i ($i = 1, 2, \ldots, \mathcal{N}$). They satisfy $\Omega^{ij}(Q^j)^c = Q^i$, where Ω^{ij} is a constant antisymmetric unitary matrix. The number $\mathcal{N}$ must be even. The anticommutation relations of the supercharges are

$$\{Q^i, Q^{jT}\} = -2\mathrm{i}\gamma^a C^T P_a \Omega^{ij} + C^T Z^{ij}, \tag{3.32}$$

where $C = C_-$, $Z^{ij} = -Z^{ji}$ for $D = 5 \bmod 8$ and $C = C_+$, $Z^{ij} = Z^{ji}$ for $D = 7 \bmod 8$. The automorphism group is $K = USp(\mathcal{N})$ and Q^i belong to the representation $\mathcal{N}$ of K.

The central charges Z^{ij} introduced above are scalar under the Lorentz transformation. More generally, one can introduce antisymmetric tensor "central charges" $Z_{ab\ldots}$ [12, 13] (Strictly speaking, they are not central since they do not commute with the Lorentz generators). We will comment on them when we discuss the commutation relations of local supertransformations in 11-dimensional supergravity in Sect. 5.2.

3.3 Supermultiplets

As we saw in Chap. 2, particle states in supersymmetric field theories in Minkowski spacetime belong to supermultiplets of the super Poincaré algebra. Similarly, if a supergravity has a Minkowski spacetime solution invariant under the super Poincaré transformations, particle states corresponding to fluctuations of the fields around the solution belong to supermultiplets of the super Poincaré algebra. As in the four-dimensional case in Chap. 2, all the states in a massless irreducible representation are obtained by applying components of the supercharges $Q_{\frac{1}{2}}$ with helicity $+\frac{1}{2}$ on the lowest helicity state $|h_0\rangle$. Here, the helicity is defined as the eigenvalue of the generator of an $SO(2)$ subgroup of $SO(D-2)$, the little group of massless particle states. Half of the components of the supercharges have helicity $+\frac{1}{2}$ while the other half have $-\frac{1}{2}$. The state $Q_{\frac{1}{2}}Q_{\frac{1}{2}} \ldots Q_{\frac{1}{2}}|h_0\rangle$ with n $Q_{\frac{1}{2}}$'s has helicity $h = h_0 + \frac{1}{2}n$. When the supercharges have too many components, all the representations contain particles with helicity $|h| > 2$. However, consistent interacting massless field theories with helicity $|h| > 2$ are not known. Hence, we shall consider the algebras which allow representations with helicity $|h| \leq 2$. There are only a finite number of possible $(D, \mathcal{N})$ for such algebras, which are listed in Table 3.3. In particular, the spacetime dimension must be $D \leq 11$. The supermultiplets for these $(D, \mathcal{N})$ were given in [11].

Supermultiplets which contain a graviton and gravitinos are called supergravity multiplets. They are massless representations of the algebra, for which $P_a P^a = 0$, $Z^{ij} = 0$. The field content of the supergravity multiplet for each $(D, \mathcal{N})$ is listed in Table 3.3. In this table $e_\mu{}^a$, ψ_μ, $B_{\mu\nu\ldots}$, λ and ϕ represent a vielbein, Rarita–Schwinger fields, antisymmetric tensor fields, spin $\frac{1}{2}$ spinor fields and scalar fields respectively. The subscripts $\pm$ on the fermionic fields denote chiralities. The superscripts $(\pm)$ on the antisymmetric tensor fields mean that they are (anti-)self-dual. The numbers of the fields are counted by real fields for the bosonic fields and by (symplectic/pseudo) Majorana(–Weyl) spinors for the fermionic fields. The last column n denotes bosonic (= fermionic) physical degrees of freedom (See below). In each dimension the theory with the largest number of $\mathcal{N}$ is called the maximal supergravity. For $D = 10$ there are two maximal supergravities: $\mathcal{N} = (1, 1)$ and $\mathcal{N} = (2, 0)$.

For $D \leq 10$ there also exist matter supermultiplets, which do not contain a graviton or gravitinos. The field contents of massless matter supermultiplets are listed in Table 3.4. In this Table A_μ, $B_{\mu\nu}^{(-)}$, χ, ψ, ϕ represent vector fields, anti self-dual antisymmetric tensor fields, spinor fields, spinor fields and scalar fields, respectively. The supermultiplets containing vector fields are called the gauge multiplets while those consisting of spinor fields and four scalar fields are called the hypermultiplets. The supermultiplets containing antisymmetric tensor fields in $D = 6$ are called the tensor multiplets.

As a check of the field contents in Tables 3.3, 3.4 one can count the numbers of bosonic and fermionic degrees of freedom in each supermultiplet, which should be the same. The physical degrees of freedom of massless fields are most easily counted in the light-cone gauge, where only transverse components of the fields are physical (See, e.g., Chap. 10 of [14]), and are given by

$$n(e_\mu{}^a) = \tfrac{1}{2}(D-2)(D-1) - 1 \,, \quad n(B_{\mu_1\ldots\mu_n}) = {}_{D-2}C_n \,, \quad n(\phi) = 1 \,,$$
$$n(\psi_\mu) = \tfrac{1}{2}(D-2-1)\cdot 2^{[D/2]} \,, \quad n(\lambda) = \tfrac{1}{2}\cdot 2^{[D/2]} \,. \tag{3.33}$$

Table 3.3 Supergravity multiplets

D	$\mathcal{N}$	Spinor	Field	n
11	1	M	$e_\mu{}^a, \psi_\mu, B_{\mu\nu\rho}$	128
10	(1,1)	MW	$e_\mu{}^a, \psi_{\mu+}, \psi_{\mu-}, B_{\mu\nu\rho}, B_{\mu\nu}, B_\mu, \lambda_+, \lambda_-, \phi$	128
	(2,0)	MW	$e_\mu{}^a, 2\psi_{\mu+}, B^{(+)}_{\mu\nu\rho\sigma}, 2B_{\mu\nu}, 2\lambda_-, 2\phi$	128
	(1,0)	MW	$e_\mu{}^a, \psi_{\mu+}, B_{\mu\nu}, \lambda_-, \phi$	64
9	2	pM	$e_\mu{}^a, 2\psi_\mu, B_{\mu\nu\rho}, 2B_{\mu\nu}, 3B_\mu, 4\lambda, 3\phi$	128
	1	pM	$e_\mu{}^a, \psi_\mu, B_{\mu\nu}, B_\mu, \lambda, \phi$	56
8	2	pM	$e_\mu{}^a, 2\psi_\mu, B_{\mu\nu\rho}, 3B_{\mu\nu}, 6B_\mu, 6\lambda, 7\phi$	128
	1	pM	$e_\mu{}^a, \psi_\mu, B_{\mu\nu}, 2B_\mu, \lambda, \phi$	48
7	4	sM	$e_\mu{}^a, 4\psi_\mu, 5B_{\mu\nu}, 10B_\mu, 16\lambda, 14\phi$	128
	2	sM	$e_\mu{}^a, 2\psi_\mu, B_{\mu\nu}, 3B_\mu, 2\lambda, \phi$	40
6	(4,4)	sMW	$e_\mu{}^a, 4\psi_{\mu+}, 4\psi_{\mu+}, 5B_{\mu\nu}, 16B_\mu, 20\lambda_+, 20\lambda_-, 25\phi$	128
	(4,2)	sMW	$e_\mu{}^a, 4\psi_{\mu+}, 2\psi_{\mu-}, 5B^{(+)}_{\mu\nu}, B^{(-)}_{\mu\nu}, 8B_\mu, 10\lambda_+, 4\lambda_-, 5\phi$	64
	(2,2)	sMW	$e_\mu{}^a, 2\psi_{\mu+}, 2\psi_{\mu-}, B_{\mu\nu}, 4B_\mu, 2\lambda_+, 2\lambda_-, \phi$	32
	(4,0)	sMW	$e_\mu{}^a, 4\psi_{\mu+}, 5B^{(+)}_{\mu\nu}$	24
	(2,0)	sMW	$e_\mu{}^a, 2\psi_{\mu+}, B^{(+)}_{\mu\nu}$	12
5	8	spM	$e_\mu{}^a, 8\psi_\mu, 27B_\mu, 48\lambda, 42\phi$	128
	6	spM	$e_\mu{}^a, 6\psi_\mu, 15B_\mu, 20\lambda, 14\phi$	64
	4	spM	$e_\mu{}^a, 4\psi_\mu, 6B_\mu, 4\lambda, \phi$	24
	2	spM	$e_\mu{}^a, 2\psi_\mu, B_\mu$	8
4	8	M	$e_\mu{}^a, 8\psi_\mu, 28B_\mu, 56\lambda, 70\phi$	128
	6	M	$e_\mu{}^a, 6\psi_\mu, 16B_\mu, 26\lambda, 30\phi$	64
	5	M	$e_\mu{}^a, 5\psi_\mu, 10B_\mu, 11\lambda, 10\phi$	32
	4	M	$e_\mu{}^a, 4\psi_\mu, 6B_\mu, 4\lambda, 2\phi$	16
	3	M	$e_\mu{}^a, 3\psi_\mu, 3B_\mu, \lambda$	8
	2	M	$e_\mu{}^a, 2\psi_\mu, B_\mu$	4
	1	M	$e_\mu{}^a, \psi_\mu$	2

The number -1 for $e_\mu{}^a$ and ψ_μ comes from the $(\gamma\text{-})$traceless condition. The factor $\frac{1}{2}$ for the fermionic fields is due to the fact that their field equations are first order differential equations. The numbers of the bosonic and fermionic degrees of freedom in each supergravity multiplet are indeed the same and are given in the last column n of the Tables. All of the maximal supergravity multiplets have 128 degrees of freedom.

3.4 Massless Sectors of M Theory and Superstring Theory

Particle states of M theory and superstring theory in Minkowski spacetime belong to supermultiplets of the super Poincaré algebra. In particular, massless states belong to supergravity multiplets and gauge multiplets, and their low energy effective theories are supergravities and super Yang–Mills theories.

Table 3.4 Massless matter multiplets

D	$\mathcal{N}$	Spinor	Field	n
10	(1,0)	MW	A_μ, χ_+	8
9	1	pM	A_μ, χ, ϕ	8
8	1	pM	$A_\mu, \chi, 2\phi$	8
7	2	sM	$A_\mu, 2\chi, 3\phi$	8
6	(2,2)	sMW	$A_\mu, 2\chi_+, 2\chi_-, 4\phi$	8
	(4,0)	sMW	$B_{\mu\nu}^{(-)}, 4\psi_-, 5\phi$	8
	(2,0)	sMW	$B_{\mu\nu}^{(-)}, 2\psi_-, \phi$	4
		sMW	$A_\mu, 2\chi_+$	4
		sMW	$2\psi_-, 4\phi$	4
5	4	spM	$A_\mu, 4\chi, 5\phi$	8
	2	spM	$A_\mu, 2\chi, \phi$	4
		spM	$2\psi, 4\phi$	4
4	4	M	$A_\mu, 4\chi, 6\phi$	8
	2	M	$A_\mu, 2\chi, 2\phi$	4
		M	$2\psi, 4\phi$	4
	1	M	A_μ, χ	2
		M	$\psi, 2\phi$	2

In the Neveu–Schwarz–Ramond formalism of superstring theory physical states of open strings are divided into the Neveu–Schwarz (NS) sector for bosonic states and the Ramond (R) sector for fermionic states. These two sectors correspond to two possible boundary conditions of the fermionic variables of strings. For closed strings, left-moving waves and right-moving waves on a string are independent and one can impose independent boundary conditions on them. Thus, physical states of closed strings are divided into four sectors: the NS–NS and R–R sectors for bosonic states and the NS–R and R–NS sectors for fermionic states.

The correspondence between M/superstring theory and supergravity is as follows.

- M Theory

M theory is an 11-dimensional theory and its low energy effective theory is $D = 11$, $\mathcal{N} = 1$ supergravity. The massless field content is

$$e_\mu{}^a, \quad B_{\mu\nu\rho}, \quad \psi_\mu. \tag{3.34}$$

- Type IIA Superstring Theory

Type IIA superstring theory is a 10-dimensional theory of closed strings and its low energy effective theory is $D = 10$, $\mathcal{N} = (1, 1)$ supergravity. The massless field content is

$$\text{NS–NS sector: } e_\mu{}^a, \quad B_{\mu\nu}, \quad \phi,$$
$$\text{R–R sector: } B_{\mu\nu\rho}, \quad B_\mu,$$
$$\text{NS–R, R–NS sectors: } \psi_{\mu+}, \quad \psi_{\mu-}, \quad \lambda_+, \quad \lambda_-. \tag{3.35}$$

- Type IIB Superstring Theory

Type IIB superstring theory is a 10-dimensional theory of closed strings and its low energy effective theory is $D = 10$, $\mathcal{N} = (2, 0)$ supergravity. The massless field content is

$$\text{NS–NS sector: } e_\mu{}^a, \quad B_{\mu\nu}^{(1)}, \quad \phi,$$
$$\text{R–R sector: } B_{\mu\nu\rho\sigma}^{(+)}, \quad B_{\mu\nu}^{(2)}, \quad C,$$
$$\text{NS–R, R–NS sectors: } 2\psi_{\mu+}, \quad 2\lambda_-. \tag{3.36}$$

Here, the superscripts (1), (2) distinguish the two second rank antisymmetric tensor fields, and C is a real scalar field.

- Type I Superstring Theory

Type I superstring theory is a 10-dimensional theory of closed and open strings. Its low energy effective theory is $D = 10$, $\mathcal{N} = (1, 0)$ supergravity coupled to super Yang–Mills theory with the gauge group $SO(32)$. The massless field content is

$$\text{closed NS–NS sector: } e_\mu{}^a, \quad \phi,$$
$$\text{closed R–R sector: } B_{\mu\nu},$$
$$\text{closed NS–R, R–NS sectors: } \psi_{\mu+}, \quad \lambda_-,$$
$$\text{open NS sector: } A_\mu,$$
$$\text{open R sector: } \chi_+. \tag{3.37}$$

- Heterotic String Theory

Heterotic string theory is a 10-dimensional theory of closed strings. Its low energy effective theory is $D = 10$, $\mathcal{N} = (1, 0)$ supergravity coupled to super Yang–Mills theory with the gauge group $SO(32)$ or $E_8 \times E_8$. This theory has fermionic variables only for the left-moving wave on a string. Physical states are divided into the NS sector and the R sector according to the boundary conditions of the fermionic variables. The massless field content is

$$\text{NS sector: } e_\mu{}^a, \quad B_{\mu\nu}, \quad \phi, \quad A_\mu,$$
$$\text{R sector: } \psi_{\mu+}, \quad \lambda_-, \quad \chi_+. \tag{3.38}$$

Table 3.5 Super anti de Sitter algebras

D	$\mathcal{N}$	Superalgebra	Bosonic subalgebra	Supercharge
7	$2, 4, 6, \ldots$	$OSp(6, 2\vert\mathcal{N})$	$SO(2, 6) \times USp(\mathcal{N})$	$(\mathbf{8}, \mathcal{N})$
6	$(2, 2)$	$F(4)$	$SO(2, 5) \times USp(2)$	$(\mathbf{8}, \mathbf{2})$
5	$2, 4, 6, 10, \ldots$	$SU(2, 2\vert\frac{1}{2}\mathcal{N})$	$SO(2, 4) \times U(\frac{1}{2}\mathcal{N})$	$(\mathbf{4}, \frac{1}{2}\mathcal{N})$
	8	$SU(2, 2\vert4)$	$SO(2, 4) \times SU(4)$	$(\mathbf{4}, \mathbf{4})$
4	$1, 2, 3, \ldots$	$OSp(\mathcal{N}\vert4)$	$SO(2, 3) \times SO(\mathcal{N})$	$(\mathbf{4}, \mathcal{N})$

3.5 Super Anti de Sitter Algebras

Some of supergravities have anti de Sitter (AdS) spacetime rather than Minkowski spacetime as a solution of the field equations. The AdS supergravities discussed in Sects. 2.5 and 2.8 are examples of such theories. Particle states corresponding to fluctuations of the fields around the anti de Sitter solution belong to supermultiplets of the super anti de Sitter (AdS) algebra. D-dimensional AdS spacetime has the isometry $SO(2, D - 1)$. The super AdS algebras are superalgebras containing the Lie algebra of $SO(2, D - 1)$. The super Poincaré algebra exists for any spacetime dimensions D and any number of supersymmetries $\mathcal{N}$. However, the super AdS algebra exists only for restricted $(D, \mathcal{N})$ [6], which we have listed for $D \geq 4$ in Table 3.5. In particular, there exits no super AdS algebra for $D \geq 8$. When an AdS spacetime solution is supersymmetric, its symmetry is one of these algebras. Therefore, there is no supersymmetric AdS spacetime solution in higher than seven dimensions. Since $SO(2, D - 1)$ is the group of conformal transformations in $D - 1$ dimensions, the superalgebras in Table 3.5 are also called the superconformal algebras.

There exist also the super de Sitter algebras, which contain the Lie algebra of the isometry $SO(1, D)$ of de Sitter space [6]. However, their representation spaces contain ghost states with a negative norm [5, 8]. Therefore, the super de Sitter algebras are not realized in a physically sensible theory.

The generators of the super AdS algebra are the $SO(2, D - 1)$ generators M_{AB} $(A, B = 0, 1, \ldots, D)$, the generators T_I of a compact group K' $(I = 1, 2, \ldots, \dim K')$, the supercharges Q^i and the central charges. The bosonic subgroup $SO(2, D-1) \times K'$ for each $(D, \mathcal{N})$ is shown in Table 3.5. The non-vanishing (anti)commutation relations of these generators are

$$[M_{AB}, M_{CE}] = -i\eta_{BC}M_{AE} + i\eta_{BE}M_{AC} + i\eta_{AC}M_{BE} - i\eta_{AE}M_{BC},$$

$$[T_I, T_J] = if_{IJ}{}^K T_K, \quad [M_{AB}, Q^i] = \frac{1}{2}i\sigma_{AB}Q^i, \quad [T_I, Q^i] = -(t_I)^i{}_j Q^j,$$

$$\{Q^i, \bar{Q}_j\} = i\sigma^{AB}M_{AB}\delta^i_j + (t^I)^i{}_j T_I. \tag{3.39}$$

Here, $\eta_{AB} = \mathrm{diag}(-1, +1, \ldots, +1, -1)$ is the $SO(2, D - 1)$ invariant metric, and σ_{AB} is defined by

$$\sigma_{ab} = \gamma_{ab}, \qquad \sigma_{aD} = \gamma_a \quad (a, b = 0, 1, \ldots, D - 1) \tag{3.40}$$

in terms of the gamma matrices γ^a of $SO(1, D-1)$. $f_{IJ}{}^K$ and t_I are the structure constant and representation matrices of the Lie algebra of K'. In the last column of Table 3.5 we listed the representations of the bosonic subalgebra $SO(2, D-1) \times K'$ that the supercharges belong to. For simplicity, we ignored the central charges in (3.39). We note that the generators T_I are not those of an automorphism group since they appear in the anticommutator of the supercharges.

AdS spacetime reduces to Minkowski spacetime in the limit of large radius. Accordingly, the super AdS algebra reduces to the super Poincaré algebra in a certain limit. To see this, we divide the generators M_{AB} into M_{ab} and M_{aD} ($a, b = 0, 1, \ldots, D-1$), and redefine the generators by introducing a real parameter m as

$$M_{aD} \to \frac{1}{m} P_a , \quad T_I \to \frac{1}{m} T_I , \quad Q^i \to \frac{1}{\sqrt{m}} Q^i . \qquad (3.41)$$

The parameter m corresponds to the inverse radius of AdS spacetime in (2.45). Using these new generators (3.39) can be rewritten as

$$[M_{ab}, M_{cd}] = -i\eta_{bc} M_{ad} + i\eta_{bd} M_{ac} + i\eta_{ac} M_{bd} - i\eta_{ad} M_{bc} ,$$

$$[M_{ab}, P_c] = -i\eta_{bc} P_a + i\eta_{ac} P_b , \quad [P_a, P_b] = -im^2 M_{ab} , \quad [T_I, T_J] = imf_{IJ}{}^K T_K ,$$

$$[M_{ab}, Q^i] = \frac{1}{2} i\gamma_{ab} Q^i , \quad [P_a, Q^i] = \frac{1}{2} im\gamma_a Q^i , \quad [T_I, Q^i] = -m(t_I)^i{}_j Q^j ,$$

$$\{Q^i, \bar{Q}_j\} = -2i\gamma^a P_a \delta^i_j + im\sigma^{ab} M_{ab}\delta^i_j + (t^I)^i{}_j T_I . \qquad (3.42)$$

In the limit $m \to 0$ we obtain the (anti)commutation relations of the super Poincaré algebra. This mathematical manipulation is called the Inönü–Wigner contraction. P_a and M_{ab} become the translation generators and the Lorentz generators. T_I commute with all the generators and are the central charges.

There is another Inönü–Wigner contraction of the super AdS algebra (3.39). If we do not make the replacement of T_I in (3.41), the (anti)commutation relations containing T_I become

$$[T_I, T_J] = if_{IJ}{}^K T_K , \quad [T_I, Q^i] = -(t_I)^i{}_j Q^j ,$$

$$\{Q^i, \bar{Q}_j\} = -2i\gamma^a P_a \delta^i_j + im\sigma^{ab} M_{ab}\delta^i_j + m(t^I)^i{}_j T_I \qquad (3.43)$$

instead of those in (3.42). In the limit $m \to 0$ we obtain the super Poincaré algebra again, but T_I in this case become the generators of a subgroup K' of the automorphism group K.

As in the case of the super Poincaré algebra we can discuss unitary representations of the super AdS algebras. For instance, unitary representations of the super AdS algebra $OSp(\mathcal{N}|4)$ for $D = 4$ were given in [1–3, 7]. When supergravity has a supersymmetric AdS solution, particle states corresponding to small fluctuations of the fields around the solution belong to such representations.

References

1. P. Breitenlohner, D.Z. Freedman, Stability in gauged extended supergravity. Ann. Phys. **144**, 249 (1982)
2. D.Z. Freedman, H. Nicolai, Multiplet shortening in $Osp(\mathcal{N}, 4)$. Nucl. Phys. **B237**, 342 (1984)
3. W. Heidenreich, All linear unitary irreducible representations of de Sitter supersymmetry with positive energy. Phys. Lett. **B110**, 461 (1982)
4. T. Kugo, P.K. Townsend, Supersymmetry and the division algebras. Nucl. Phys. **B221**, 357 (1983)
5. J. Lukierski, A. Nowicki, All possible de Sitter superalgebras and the presence of ghosts. Phys. Lett. **B151**, 382 (1985)
6. W. Nahm, Supersymmetries and their representations. Nucl. Phys. **B135**, 149 (1978)
7. H. Nicolai, Representations of supersymmetry in anti-de Sitter space, in *Supersymmetry and Supergravity '84*, ed. by B. de Wit, P. Fayet, P. van Nieuwenhuizen (World Scientific, Singapore, 1984)
8. K. Pilch, P. van Nieuwenhuizen, M.F. Sohnius, De Sitter superalgebras and supergravity. Commun. Math. Phys. **98**, 105 (1985)
9. J.J. Sakurai, *Advanced Quantum Mechanics*, (Addison-Wesley, Reading, 1967)
10. J. Scherk, Extended supersymmetry and extended supergravity theories, in *Recent Developments in Gravitation*, ed. by M. Lévy, S. Deser (Plenum Press, New York, 1979)
11. J. Strathdee, Extended Poincaré supersymmetry. Int. J. Mod. Phys. **A2**, 273 (1987)
12. P.K. Townsend, "*P*-brane democracy", in *Particles, Strings and Cosmology*, ed. by J. Bagger, G. Domokos, A. Falk, S. Kovesi-Domokos (World Scientific, Singapore, 1996). [hep-th/9507048]
13. J.W. van Holten, A. van Proeyen, $\mathcal{N} = 1$ supersymmetry algebras in $D = 2$, $D = 3$, $D = 4$ mod 8. J. Phys. **A15**, 3763 (1982)
14. B. Zwiebach, *A First Course in String Theory*, 2nd edn. (Cambridge University Press, Cambridge, 2009)

Chapter 4
Global Non-compact Symmetries

4.1 Non-linear Sigma Models

Scalar fields appearing in supergravities are described by a G/H non-linear sigma model, where G is a non-compact Lie group and H is a maximal compact subgroup of G. The G/H non-linear sigma model is a theory of G/H-valued scalar fields, which is invariant under non-linearly realized global G transformations. Although the Killing form of the non-compact Lie group G is not positive definite, the coset space G/H by its maximal compact subgroup H has a positive definite metric. Using this fact we can construct a theory of scalar fields which does not contain ghost fields with a kinetic term of wrong sign. In this section we shall review how to construct the G/H non-linear sigma model [1, 2].

We represent the scalar fields by a G-valued scalar field $V(x)$ and require the local H invariance. Since we do not introduce independent H gauge fields, the H part of $V(x)$ can be gauged away and its physical degrees of freedom are on the coset space G/H. The number of independent scalar fields is $\dim G/H = \dim G - \dim H$. The global G acts on $V(x)$ from the left and the local H from the right as

$$V(x) \rightarrow g V(x) h^{-1}(x) \qquad (g \in G, \ h(x) \in H). \tag{4.1}$$

We construct a Lagrangian invariant under these transformations.

We decompose the Lie algebra $\mathscr{G}$ of G as $\mathscr{G} = \mathscr{H} + \mathscr{N}$, where $\mathscr{H}$ is the Lie algebra of H and $\mathscr{N}$ is its orthogonal complement in $\mathscr{G}$. The orthogonality is defined by the Killing form of $\mathscr{G}$. Using the trace over representation matrices the orthogonality implies $\mathrm{tr}(\mathscr{H}\mathscr{N}) = 0$. It can be easily shown that

$$[\mathscr{H}, \mathscr{H}] \subset \mathscr{H}, \qquad [\mathscr{H}, \mathscr{N}] \subset \mathscr{N}. \tag{4.2}$$

The $\mathscr{G}$-valued field $V^{-1}\partial_\mu V$ is decomposed as

$$V^{-1}\partial_\mu V = Q_\mu + P_\mu, \qquad Q_\mu \in \mathscr{H}, \quad P_\mu \in \mathscr{N}. \tag{4.3}$$

By (4.2), Q_μ and P_μ transform under the local H in (4.1) as

Y. Tanii, *Introduction to Supergravity*, SpringerBriefs in Mathematical Physics, DOI: 10.1007/978-4-431-54828-7_4, © The Author(s) 2014

$$Q_\mu \to h Q_\mu h^{-1} + h \partial_\mu h^{-1}, \qquad P_\mu \to h P_\mu h^{-1}, \tag{4.4}$$

while they are invariant under the global G in (4.1). We see that Q_μ transforms as an H gauge field while P_μ is covariant under H. By (4.3), P_μ can be expressed as $P_\mu = V^{-1} \left(\partial_\mu V - V Q_\mu \right) \equiv V^{-1} D_\mu V$, where D_μ is the H-covariant derivative on V.

By using these quantities we can construct a Lagrangian which is invariant under the global G and the local H transformations. The kinetic term of the scalar fields is

$$\mathcal{L} = -a\, e \operatorname{tr}(P_\mu P^\mu), \tag{4.5}$$

where a is a positive constant. This Lagrangian is quadratic in the derivative of V and is manifestly invariant under the global G and the local H transformations. The H gauge field Q_μ can be used to define the covariant derivatives on other fields which transform under the local H. For instance, when a spinor field $\psi(x)$ transforms under the local H transformation as $\psi(x) \to h(x)\psi(x)$, the covariant derivative is

$$D_\mu \psi = \left(\partial_\mu + Q_\mu \right) \psi. \tag{4.6}$$

We can also use P_μ to construct H-invariant terms in the Lagrangian such as

$$\mathcal{L} = e \bar{\psi} \gamma^\mu P_\mu \psi. \tag{4.7}$$

Furthermore, by using V we can construct invariant kinetic terms of fields which transform under the global G. See (4.15) below.

We can describe the theory in terms of the physical fields only by fixing a gauge for the local H symmetry. For instance, we can choose a gauge

$$V(x) = e^{\Phi(x)}, \tag{4.8}$$

where $\Phi(x)$ is an $\mathcal{N}$-valued field, which represents physical degrees of freedom. The global G transformation in (4.1) does not preserve this gauge condition. To preserve the gauge, the transformation $g \in G$ must be accompanied by a compensating H transformation $h(x; g)$. Therefore, the G transformation of $\Phi(x)$ is given by

$$e^{\Phi(x)} \to e^{\Phi'(x)} = g\, e^{\Phi(x)} h^{-1}(x; g), \tag{4.9}$$

where the compensating H transformation $h(x; g)$ is chosen such that $\Phi'(x)$ belongs to $\mathcal{N}$. The transformation $\Phi(x) \to \Phi'(x)$ is a non-linear realization of G. When $g \in H \subset G$, we can take $h(x; g) = g$ and g is linearly realized.

The groups G and H appearing in supergravities in Table 3.3 are listed in Table 4.1. (See Appendix A and [6] for the definitions of the groups in this Table.) We see that the dimension of the coset space G/H given in the last column is equal to the number of the scalar fields of each theory in Table 3.3. When G is

Table 4.1 G and H in supergravities

D	$\mathcal{N}$	G	H	$\dim G/H$
11	1	1	1	0
10	(1, 1)	$\mathbb{R}^+$	1	1
	(2, 0)	$SL(2, \mathbb{R})$	$SO(2)$	2
	(1, 0)	$\mathbb{R}^+$	1	1
9	2	$GL(2, \mathbb{R})$	$SO(2)$	3
	1	$\mathbb{R}^+$	1	1
8	2	$SL(3, \mathbb{R}) \times SL(2, \mathbb{R})$	$SO(3) \times SO(2)$	7
	1	$\mathbb{R}^+$	1	1
7	4	$SL(5, \mathbb{R})$	$SO(5)$	14
	2	$\mathbb{R}^+$	1	1
6	(4, 4)	$SO(5, 5)$	$SO(5) \times SO(5)$	25
	(4, 2)	$SO(5, 1)$	$SO(5)$	5
	(2, 2)	$\mathbb{R}^+$	1	1
	(4, 0)	$USp(4)$	$USp(4)$	0
	(2, 0)	$USp(2)$	$USp(2)$	0
5	8	$E_{6(+6)}$	$USp(8)$	42
	6	$SU^*(6)$	$USp(6)$	14
	4	$USp(4) \times \mathbb{R}^+$	$USp(4)$	1
	2	$USp(2)$	$USp(2)$	0
4	8	$E_{7(+7)}$	$SU(8)$	70
	6	$SO^*(12)$	$U(6)$	30
	5	$SU(5, 1)$	$U(5)$	10
	4	$SU(4) \times SL(2, \mathbb{R})$	$U(4)$	2
	3	$U(3)$	$U(3)$	0
	2	$U(2)$	$U(2)$	0
	1	$U(1)$	$U(1)$	0

compact, $H = G$, and the theory has no scalar fields. Among the fields contained in supergravities the bosonic fields other than the gravitational field and the scalar fields transform under only G, while the fermionic fields transform under only H. The gravitational field is invariant under both of G and H. The scalar fields transform under both of G and H.

4.1.1 $SL(2, \mathbb{R})/SO(2)$ *Non-linear Sigma Model*

As an example of G/H non-linear sigma models let us consider the case $G = SL(2, \mathbb{R}) \sim SU(1, 1)$ and $H = SO(2) \sim U(1)$. This sigma model appears in $D = 10$, $\mathcal{N} = (2, 0)$ and $D = 4$, $\mathcal{N} = 4$ supergravities. The $SL(2, \mathbb{R})$-valued scalar field is represented by a real 2×2 matrix $V(x)$ with $\det V = 1$. The global $SL(2, \mathbb{R})$ transformation and the local $SO(2)$ transformation of $V(x)$ are given by (4.1) with

$$g = \begin{pmatrix} a & b \\ c & d \end{pmatrix} \qquad (a, b, c, d \in \mathbb{R}, \; ad - bc = 1),$$

$$h(x) = \begin{pmatrix} \cos\theta(x) & -\sin\theta(x) \\ \sin\theta(x) & \cos\theta(x) \end{pmatrix} \qquad (\theta(x) \in \mathbb{R}). \tag{4.10}$$

We can fix a gauge for the local $SO(2)$ by choosing V as

$$V(x) = \begin{pmatrix} e^{-\frac{1}{2}\phi(x)} & C(x)e^{\frac{1}{2}\phi(x)} \\ 0 & e^{\frac{1}{2}\phi(x)} \end{pmatrix}, \tag{4.11}$$

where $\phi(x)$ and $C(x)$ are real scalar fields. P_μ and Q_μ in (4.3) are the symmetric part and the antisymmetric part of $V^{-1}\partial_\mu V$, respectively, and are given in this gauge by

$$P_\mu = \frac{1}{2} \begin{pmatrix} -\partial_\mu\phi & e^\phi\partial_\mu C \\ e^\phi\partial_\mu C & \partial_\mu\phi \end{pmatrix}, \quad Q_\mu = -\frac{1}{2}e^\phi\partial_\mu C \begin{pmatrix} 0 & 1 \\ -1 & 0 \end{pmatrix}. \tag{4.12}$$

The Lagrangian (4.5) with $a = 1$ then becomes

$$\begin{aligned}
\mathcal{L} &= -e\,\mathrm{tr}\left(P_\mu P^\mu\right) \\
&= -\frac{1}{2}e\left(\partial_\mu\phi\partial^\mu\phi + e^{2\phi}\partial_\mu C\partial^\mu C\right) \\
&= -\frac{1}{2}e\,\frac{\partial_\mu\tau^*\partial^\mu\tau}{(\mathrm{Im}\,\tau)^2},
\end{aligned} \tag{4.13}$$

where we have introduced a complex scalar field taking values in the upper half complex plane $\tau(x) = C(x) + \mathrm{i}\,e^{-\phi(x)}$. The global $SL(2, \mathbb{R})$ acts on this complex field as a linear fractional transformation

$$\tau \to \frac{a\tau + b}{c\tau + d}. \tag{4.14}$$

It is easy to see that the Lagrangian (4.13) is indeed invariant under (4.14).

As an example of couplings to other fields which transform under G, let us consider a couple of vector fields $\mathscr{A}_\mu = (A_\mu^{(1)}, A_\mu^{(2)})$ with the Lagrangian

$$\mathcal{L} = -\frac{1}{4}e\mathscr{F}_{\mu\nu}VV^T\mathscr{F}^{T\mu\nu}, \tag{4.15}$$

where $\mathscr{F}_{\mu\nu} = 2\partial_{[\mu}\mathscr{A}_{\nu]}$ is the field strength. If the vector fields transform as $\mathscr{A}_\mu \to \mathscr{A}_\mu g^{-1}$ under the $SL(2, \mathbb{R})$, this Lagrangian is invariant under the global $SL(2, \mathbb{R})$ and the local $SO(2)$. Although $\mathscr{A}_\mu$ transforms in a representation of the non-compact group $SL(2, \mathbb{R})$, this theory does not contain ghost fields which have kinetic terms of wrong sign.

4.2 Duality Symmetry

In some of supergravities in even dimensions the global symmetry G in Table 4.1 is realized as a duality symmetry, which we discuss in this section. The duality symmetry is a generalization of the electric–magnetic duality of Maxwell's equations.

Let us first consider free Maxwell's equations as a simple example in order to explain what the duality symmetry is. When there is no charges or currents, Maxwell's equations consist of the field equation and the Bianchi identity

$$\partial_\mu(e F^{\mu\nu}) = 0, \qquad \partial_\mu(e * F^{\mu\nu}) = 0, \tag{4.16}$$

where the field strength $F_{\mu\nu}$ and its Hodge dual $*F^{\mu\nu}$ are defined by

$$F_{\mu\nu} = \partial_\mu A_\nu - \partial_\nu A_\mu, \qquad *F^{\mu\nu} = \frac{1}{2} e^{-1} \varepsilon^{\mu\nu\rho\sigma} F_{\rho\sigma}. \tag{4.17}$$

Equations (4.16) are invariant under the global transformation

$$\delta \begin{pmatrix} F^{\mu\nu} \\ *F^{\mu\nu} \end{pmatrix} = \begin{pmatrix} A & B \\ C & D \end{pmatrix} \begin{pmatrix} F^{\mu\nu} \\ *F^{\mu\nu} \end{pmatrix}, \tag{4.18}$$

where A, B, C, D are infinitesimal real constant parameters. Here, we have to take account of the fact that $F_{\mu\nu}$ and $*F^{\mu\nu}$ are not independent but are related by the second equation in (4.17). By taking the Hodge dual of the upper equation of (4.18) and using the identity $*^2 F = -F$, we obtain $\delta *F = A*F - BF$, which should coincide with the lower equation. Therefore, the transformation parameters must satisfy $D = A$ and $C = -B$, and (4.18) can be written as

$$\delta \begin{pmatrix} F + \mathrm{i} *F \\ F - \mathrm{i} *F \end{pmatrix} = \begin{pmatrix} A - \mathrm{i}B & 0 \\ 0 & A + \mathrm{i}B \end{pmatrix} \begin{pmatrix} F + \mathrm{i} *F \\ F - \mathrm{i} *F \end{pmatrix}. \tag{4.19}$$

We see that the transformations form the group $GL(1, \mathbb{C})$. Such a transformation, which exchanges the field equation and the Bianchi identity, is called the duality transformation. The symmetry under the duality transformation is called the duality symmetry.

The duality transformation is consistent only on-shell. Since the independent field of the theory is A_μ, $\delta F_{\mu\nu}$ in (4.19) should be derived from δA_μ as $\partial_\mu \delta A_\nu - \partial_\nu \delta A_\mu = A F_{\mu\nu} + B*F_{\mu\nu}$. The integrability of this equation requires the field equation $\partial_\mu(e F^{\mu\nu}) = 0$. Therefore, to construct a theory invariant under the duality transformation we have to study the invariance of the field equations.

4.2.1 Duality Symmetry in General Even Dimensions

Let us study the duality symmetry in general even dimensions $D = 2n$ [3, 5, 7, 8]. We consider a theory of $(n - 1)$-th rank antisymmetric tensor fields $B^I_{\mu_1\cdots\mu_{n-1}}(x)$ $(I = 1, \ldots, M)$ interacting with other fields $\varphi_i(x)$ $(i = 1, 2, 3, \ldots)$. The field strengths and their Hodge duals are defined by

$$F^I_{\mu_1\cdots\mu_n} = n\,\partial_{[\mu_1}B^I_{\mu_2\cdots\mu_n]}, \quad *F^{I\mu_1\cdots\mu_n} = \frac{1}{n!}e^{-1}\varepsilon^{\mu_1\cdots\mu_n\nu_1\cdots\nu_n}F^I_{\nu_1\cdots\nu_n}. \tag{4.20}$$

In D dimensions the Hodge dual $*$ satisfies

$$*^2F = \varepsilon F, \quad \varepsilon = \begin{cases} +1 & (D = 4k + 2) & (k = 0, 1, 2, \ldots), \\ -1 & (D = 4k) & (k = 1, 2, 3, \ldots). \end{cases} \tag{4.21}$$

We assume that the Lagrangian depends on the fields $B^I_{\mu_1\cdots\mu_{n-1}}$ only through their field strengths $F^I_{\mu_1\cdots\mu_n}$, and contains terms up to quadratic order in the field strengths. We require the duality symmetry in this theory and obtain conditions on the Lagrangian and a possible duality symmetry group.

The field equations of $B^I_{\mu_1\cdots\mu_{n-1}}$ and the Bianchi identities are

$$\partial_{\mu_1}(e*G^{\mu_1\cdots\mu_n}_I) = 0, \quad \partial_{\mu_1}(e*F^{I\mu_1\cdots\mu_n}) = 0, \tag{4.22}$$

where the antisymmetric tensors $G_{I\mu_1\cdots\mu_n}$ are defined by

$$*G^{\mu_1\cdots\mu_n}_I = -\frac{\varepsilon n!}{e}\frac{\partial\mathscr{L}}{\partial F^I_{\mu_1\cdots\mu_n}}. \tag{4.23}$$

(For a free theory, we have $G^{\mu_1\cdots\mu_n}_I = *F^{I\mu_1\cdots\mu_n}$.) Equations (4.22) are invariant under the global transformation

$$\delta\begin{pmatrix} F \\ G \end{pmatrix} = \begin{pmatrix} A & B \\ C & D \end{pmatrix}\begin{pmatrix} F \\ G \end{pmatrix}, \quad \delta\varphi^i = \xi^i(\varphi), \tag{4.24}$$

where A, B, C, D are constant $n \times n$ real matrices and $\xi^i(\varphi)$ are functions of φ^i. As in the case of Maxwell's equations, these constant matrices are not independent. We shall obtain the conditions that these matrices should satisfy by studying (i) the covariance of the definition of G_I (4.23) and (ii) the covariance of the field equations of φ^i under the transformation (4.24).

Let us first study the covariance of the definition of G_I. By (4.23), G_I are expressed in terms of F^I and φ^i. Therefore, the transformation of G_I can be derived from those of F^I and φ^i. From (4.23) we obtain

$$\delta * G_I = -\frac{\varepsilon n!}{e}\left((AF+BG)^J\frac{\partial}{\partial F^J} + \xi^i\frac{\partial}{\partial\varphi^i} + \partial_\mu\xi^i\frac{\partial}{\partial(\partial_\mu\varphi^i)}\right)\frac{\partial\mathscr{L}}{\partial F^I}$$

$$= -\frac{\varepsilon n!}{e}\frac{\partial\delta\mathscr{L}}{\partial F^I} - *G_J A^J{}_I - *G_J B^{JK}\frac{\partial G_K}{\partial F^I}. \tag{4.25}$$

This should coincide with the transformation given in the lower equation of (4.24). By equating them we obtain

$$\frac{\partial}{\partial F^I}\left(n!\delta\mathscr{L} + \frac{1}{2}\varepsilon e F^J C_{JK} * F^K + \frac{1}{2}\varepsilon e * G_J B^{JK} G_K\right) - \left(A^J{}_I + D_I{}^J\right)n!\frac{\partial\mathscr{L}}{\partial F^J}$$

$$= -\frac{1}{2}\varepsilon e\left(C_{IJ} + \varepsilon C_{JI}\right)*F^J - \frac{1}{2}\varepsilon e * G_J\left(B^{JK} + \varepsilon B^{KJ}\right)\frac{\partial G_K}{\partial F^I}. \tag{4.26}$$

When there exist nontrivial interactions, this equation gives the conditions on the transformation parameters

$$A^I{}_J + D_J{}^I = \alpha\delta_J^I, \quad B^{IJ} = -\varepsilon B^{JI}, \quad C_{IJ} = -\varepsilon C_{JI}, \tag{4.27}$$

where α is an arbitrary constant, and the condition on the variation of the Lagrangian

$$\frac{\partial}{\partial F^I}\left(\delta\mathscr{L} + \frac{1}{2n!}\varepsilon e F^J C_{JK} * F^K + \frac{1}{2n!}\varepsilon e * G_J B^{JK} G_K - \alpha\mathscr{L}\right) = 0. \tag{4.28}$$

Next, let us consider the covariance of the field equations of φ^i.

$$E_i \equiv \frac{\partial\mathscr{L}}{\partial\varphi^i} - \partial_\mu\left(\frac{\partial\mathscr{L}}{\partial(\partial_\mu\varphi^i)}\right) = 0. \tag{4.29}$$

By calculations similar to (4.25) we find that they transform as

$$\delta E_i = \left(\frac{\partial}{\partial\varphi^i} - \partial_\mu\frac{\partial}{\partial(\partial_\mu\varphi^i)}\right)\left(\delta\mathscr{L} + \frac{1}{2n!}\varepsilon e * G_I B^{IJ} G_J\right) - \frac{\partial\xi^j}{\partial\varphi^i}E_j. \tag{4.30}$$

Requiring the covariance $\delta E_i = -\frac{\partial\xi^j}{\partial\varphi^i}E_j$ we obtain another condition on the variation of the Lagrangian

$$\left(\frac{\partial}{\partial\varphi^i} - \partial_\mu\frac{\partial}{\partial(\partial_\mu\varphi^i)}\right)\left(\delta\mathscr{L} + \frac{1}{2n!}\varepsilon e * G_I B^{IJ} G_J\right) = 0. \tag{4.31}$$

Now we can find out a possible duality group by studying (4.27), (4.28) and (4.31). Comparing (4.28) and (4.31) we find $\alpha = 0$. Then, the conditions on the parameters (4.27) can be written as

$$X^T \Omega + \Omega X = 0, \quad X = \begin{pmatrix} A & B \\ C & D \end{pmatrix}, \quad \Omega = \begin{pmatrix} 0 & \varepsilon \\ 1 & 0 \end{pmatrix}. \qquad (4.32)$$

For $D = 4k$ ($\varepsilon = -1$), Ω is an antisymmetric matrix and this condition implies that the group of duality transformations is $Sp(2M, \mathbb{R})$ or its subgroup. On the other hand, for $D = 4k + 2$ ($\varepsilon = +1$), Ω is a symmetric matrix, which can be diagonalized to $\mathrm{diag}(1, -1)$. Therefore, the group of duality transformations in this case is $SO(M, M)$ or its subgroup. Equations (4.27), (4.28) and (4.31) also restrict the variation of the Lagrangian as

$$\delta\mathscr{L} = -\frac{1}{2n!}\varepsilon e F^I C_{IJ} * F^J - \frac{1}{2n!}\varepsilon e * G_I B^{IJ} G_J = \delta\left(-\frac{1}{2n!}\varepsilon e F^I * G_I\right). \quad (4.33)$$

Thus, although the Lagrangian is not invariant under the duality transformation, it transforms in a definite way.

It can be shown that a derivative of the Lagrangian with respect to a parameter λ of the theory is invariant under the duality transformation. Here, we assume that λ is invariant under the duality transformation and that ξ^i in (4.24) do not depend on λ. To show this we compute the derivative of $\delta\mathscr{L}$ with respect to λ as

$$\frac{\partial}{\partial\lambda}\delta\mathscr{L} = \frac{\partial}{\partial\lambda}\left[(AF + BG)^I\frac{\partial\mathscr{L}}{\partial F^I} + \xi^i\frac{\partial\mathscr{L}}{\partial\varphi^i} + \partial_\mu\xi^i\frac{\partial\mathscr{L}}{\partial(\partial_\mu\varphi^i)}\right]$$
$$= \delta\left(\frac{\partial\mathscr{L}}{\partial\lambda}\right) - \frac{\varepsilon}{n!}e * G_I B^{IJ}\frac{\partial G_J}{\partial\lambda}. \qquad (4.34)$$

Therefore, we obtain

$$\delta\left(\frac{\partial\mathscr{L}}{\partial\lambda}\right) = \frac{\partial}{\partial\lambda}\left(\delta\mathscr{L} + \frac{1}{2n!}\varepsilon e F^I C_{IJ} * F^J + \frac{1}{2n!}\varepsilon e * G_I B^{IJ} G_J\right), \quad (4.35)$$

which vanishes by (4.33). The parameter λ can be an external field invariant under the duality transformation such as the metric. Thus, the energy–momentum tensor obtained as the derivative of the Lagrangian with respect to the metric is invariant under the duality transformation.

Let us obtain the explicit form of the Lagrangian which transforms as in (4.33). It can be written as

$$\mathscr{L} = -\frac{1}{2n!}\varepsilon e F^I * G_I + \text{(invariant terms)}$$
$$= -\frac{1}{2n!}\varepsilon e F^I * G_I - \frac{1}{2n!}\varepsilon e (F^I I_I + \varepsilon G_I H^I) + \mathscr{L}_{\mathrm{inv}}(\varphi, \partial\varphi), \qquad (4.36)$$

where the n-th rank antisymmetric tensors $(H^I_{\mu_1\cdots\mu_n}, I_{I\,\mu_1\cdots\mu_n})$ are functions of φ^i and $\partial_\mu\varphi^i$ which transform in the same way as $(F^I_{\mu_1\cdots\mu_n}, G_{I\,\mu_1\cdots\mu_n})$ under (4.24), and $\mathscr{L}_{\mathrm{inv}}(\varphi, \partial\varphi)$ is invariant under (4.24). In the second line we have assumed that the duality symmetry group is the maximal one, i.e., $Sp(2M, \mathbb{R})$ in $D = 4k$ or $SO(M, M)$ in $D = 4k + 2$, for simplicity. When the duality symmetry group is a subgroup of them, there can be other invariants than $F^I I_I + \varepsilon G_I H^I$. Substituting (4.36) into (4.23) we obtain a differential equation for $*G_I$

$$(*G - I)_I = (F - \varepsilon *H)^J \frac{\partial}{\partial F^I}(*G - I)_J . \tag{4.37}$$

The solution of this equation can be written as

$$* G_I = I_I + \varepsilon K_{IJ}(\varphi)(F^J - \varepsilon *H^J), \tag{4.38}$$

where

$$K_{IJ}(\varphi) = K_{1IJ}(\varphi) + K_{2IJ}(\varphi) * \qquad (K_1^T = K_1, \ K_2^T = -\varepsilon K_2). \tag{4.39}$$

The Hodge dual $*$ acts on antisymmetric tensors on the right of it. By the covariance of (4.38) under the duality transformation, K must transform as

$$\delta K = -KA - KBK * + \varepsilon C * + DK. \tag{4.40}$$

Substituting the solution (4.38) into (4.36) we obtain

$$\mathcal{L} = -\frac{1}{2n!}e F^I K_{IJ} F^J - \frac{1}{n!}\varepsilon e F^I (I_I - K_{IJ} *H^J)$$
$$+ \frac{1}{2n!}e *H^I (I_I - K_{IJ} *H^J) + \mathcal{L}_{\text{inv}}(\varphi, \partial\varphi). \tag{4.41}$$

Thus, if we can find out the functions $H^I(\phi, \partial\phi)$, $I_I(\phi, \partial\phi)$, $K_{IJ}(\phi)$ with the appropriate transformation properties, we have the explicit form of the Lagrangian.

4.2.2 Compact Duality Symmetry

Let us consider a special case of $K = 1$. In this case the duality symmetry group must be a compact group as we will see below. From (4.40) the transformation parameters must satisfy $A = D$, $B = \varepsilon C$. For $D = 4k$ ($\varepsilon = -1$) these conditions together with (4.32) imply $A^T = -A$, $B^T = B$ and the transformation law becomes

$$\delta \begin{pmatrix} F + iG \\ F - iG \end{pmatrix} = \begin{pmatrix} A - iB & 0 \\ 0 & A + iB \end{pmatrix} \begin{pmatrix} F + iG \\ F - iG \end{pmatrix}. \tag{4.42}$$

Since $A - iB$ is anti-hermitian, the duality symmetry group is $U(M)$ or its subgroup. Note that $U(M)$ is a maximal compact subgroup of $Sp(2M, \mathbb{R})$. On the other hand, for $D = 4k + 2$ ($\varepsilon = +1$) the above conditions imply $A^T = -A$, $B^T = -B$ and the transformation law becomes

$$\delta \begin{pmatrix} F + G \\ F - G \end{pmatrix} = \begin{pmatrix} A + B & 0 \\ 0 & A - B \end{pmatrix} \begin{pmatrix} F + G \\ F - G \end{pmatrix}. \tag{4.43}$$

Since $A + B$ and $A - B$ are real and antisymmetric, the duality symmetry group is $SO(M) \times SO(M)$ or its subgroup. Again, $SO(M) \times SO(M)$ is a maximal compact subgroup of $SO(M, M)$.

$D = 4$, $\mathcal{N} = 2$ Poincaré Supergravity

As an example of theories with a compact duality symmetry let us consider the $D = 4$, $\mathcal{N} = 2$ Poincaré supergravity discussed in Sect. 2.7. The field equations derived from the Lagrangian (2.51) are invariant under the global $U(1)$ transformation (2.61). This $U(1)$ acts on the vector field as the duality transformation. $G_{\mu\nu}$ in (4.23) can be obtained from (2.51) as

$$*G_{\mu\nu} = -F_{\mu\nu} - *H_{\mu\nu} + I_{\mu\nu}, \tag{4.44}$$

where

$$H_{\mu\nu} = -\frac{1}{2} i \varepsilon^{ij} \bar{\psi}^i_\mu \gamma_5 \psi^j_\nu, \qquad I_{\mu\nu} = \frac{1}{2} \varepsilon^{ij} \bar{\psi}^i_\mu \psi^j_\nu. \tag{4.45}$$

We can easily see that (H, I) transform in the same way as (F, G) under (2.61). Furthermore, the transformation of $G_{\mu\nu}$ in (4.44) derived from $\delta F_{\mu\nu}$ and $\delta \psi_\mu$ correctly reproduces $\delta G_{\mu\nu}$ in (2.61). Using $G_{\mu\nu}$ in (4.44), the Lagrangian (2.51) can be rewritten in the form (4.41) with $K = 1$.

4.2.3 Non-compact Duality Symmetry

We can construct $K_{IJ}(\varphi)$ which transforms as in (4.40) by using the G/H nonlinear sigma model discussed in Sect. 4.1. Here, G is the duality symmetry group, which we assume to be the maximal one, i.e., $Sp(2M, \mathbb{R})$ or $SO(M, M)$, and H is a maximal compact subgroup of G, i.e., $U(M)$ or $SO(M) \times SO(M)$.

$D = 4k$

Let us first discuss the case $D = 4k$, $G = Sp(2M, \mathbb{R})$, $H = U(M)$. In this case it is convenient to use a complex basis as in (4.42)

$$\mathscr{F}(x) = \begin{pmatrix} F(x) + iG(x) \\ F(x) - iG(x) \end{pmatrix}. \tag{4.46}$$

In this basis the G-valued scalar field has the form

$$V(x) = \begin{pmatrix} u(x) & v(x)^* \\ v(x) & u(x)^* \end{pmatrix}, \quad u^\dagger u - v^\dagger v = 1, \quad u^T v = v^T u, \quad (4.47)$$

where u and v are complex $M \times M$ matrices. Under the global G and the local H they transform as

$$\mathscr{F}(x) \rightarrow g\mathscr{F}(x), \quad V(x) \rightarrow gV(x)h(x)^{-1}, \quad (4.48)$$

where the matrices $g \in G$ and $h(x) \in H$ are

$$g = \begin{pmatrix} a & b^* \\ b & a^* \end{pmatrix}, \quad a^\dagger a - b^\dagger b = 1, \quad a^T b = b^T a,$$

$$h(x) = \begin{pmatrix} a_H(x) & 0 \\ 0 & a_H^*(x) \end{pmatrix}, \quad a_H^\dagger a_H = 1 \quad (4.49)$$

with complex $M \times M$ matrices a, b and a_H.

Using the scalar fields u and v we can construct K transforming as in (4.40) as

$$K = (u^\dagger + v^\dagger)^{-1}(u^\dagger - v^\dagger). \quad (4.50)$$

In (4.50) the imaginary unit i in complex u, v is replaced by the Hodge dual operation $*$, e.g., $u = \text{Re}\,u + \text{Im}\,u\,*$, $u^\dagger = \text{Re}\,u^T - \text{Im}\,u^T\,*$. Since the Hodge dual $*$ satisfies $*^2 = -1$ for $D = 4k$ in the same way as the imaginary unit i satisfies $i^2 = -1$, we can consistently replace i by $*$. Using (4.48) we can show that this K is H invariant and transforms as in (4.40) under G. The expression (4.50) can be simplified by introducing the field

$$z = v^*(u^*)^{-1} = z^T, \quad (4.51)$$

where the second equality can be shown by using (4.47). This field $z(x)$ is H invariant and represents physical degrees of freedom of the scalar fields. Under G it transforms as

$$z \rightarrow (az + b^*)(bz + a^*)^{-1}. \quad (4.52)$$

In terms of z the matrix K in (4.50) can be written as

$$K = \frac{1-z}{1+z} = K^T. \quad (4.53)$$

$D = 4k + 2$

We now turn to the case $D = 4k + 2$, $G = SO(M, M)$, $H = SO(M) \times SO(M)$. It is convenient to use the Ω-diagonal basis as in (4.43)

$$\mathscr{F}(x) = \begin{pmatrix} F(x) + G(x) \\ F(x) - G(x) \end{pmatrix}. \quad (4.54)$$

In this basis the G-valued scalar field has the form

$$V(x) = \begin{pmatrix} u_1(x) & v_2(x) \\ v_1(x) & u_2(x) \end{pmatrix}, \quad u_1^T u_1 - v_1^T v_1 = 1 = u_2^T u_2 - v_2^T v_2, \quad u_1^T v_2 = v_1^T u_2,$$

$$(4.55)$$

where $u_1(x)$, $v_1(x)$, $u_2(x)$, $v_2(x)$ are real $M \times M$ matrices. Under the global G and the local H they transform as in (4.48), where the matrices $g \in G$ and $h(x) \in H$ are

$$g = \begin{pmatrix} a & b \\ c & d \end{pmatrix}, \quad a^T a - c^T c = 1 = d^T d - b^T b, \quad a^T b = c^T d,$$

$$h(x) = \begin{pmatrix} a_H(x) & 0 \\ 0 & d_H(x) \end{pmatrix}, \quad a_H^T a_H = 1 = d_H^T d_H \tag{4.56}$$

with real $M \times M$ matrices a, b, c, d, a_H and d_H.

In this case K transforming as in (4.40) is

$$K = (u_1 - v_1)(u_1 + v_1)^{-1} P_+ + (u_2 - v_2)(u_2 + v_2)^{-1} P_-. \tag{4.57}$$

Since the Hodge dual $*$ satisfies $*^2 = +1$ for $D = 4k + 2$, we have introduced the projection operators $P_\pm = \frac{1}{2}(1 \pm *)$. Using (4.48) one can show that this K is H invariant and transforms as in (4.40) under G. The expression (4.57) can be simplified by introducing the field

$$z = (v_1(u_1)^{-1})^T = v_2(u_2)^{-1}, \tag{4.58}$$

where the second equality can be shown by using (4.55). This $z(x)$ is H invariant and represents physical degrees of freedom of the scalar fields. Under G it transforms as

$$z \rightarrow (az + b)(cz + d)^{-1}. \tag{4.59}$$

In terms of z the matrix K can be written as

$$K = \frac{1 - z^T}{1 + z^T} P_+ + \frac{1 - z}{1 + z} P_-. \tag{4.60}$$

4.3 $D = 4$, $\mathcal{N} = 8$ Poincaré Supergravity

As an example of supergravities with the non-compact duality symmetry let us discuss $D = 4$, $\mathcal{N} = 8$ Poincaré supergravity, which has a maximal supersymmetry in four dimensions. This theory was constructed first in [3] by a dimensional reduction from $D = 11$, $\mathcal{N} = 1$ supergravity. It was shown there that this theory has global $E_{7(+7)}$ and local $SU(8)$ symmetries. $E_{7(+7)}$ is a non-compact group and $SU(8)$ is its maximal compact subgroup. The scalar fields are described by the $E_{7(+7)}/SU(8)$

non-linear sigma model. $E_{7(+7)}$ acts on the vector fields as the duality transformation. Later, this theory was reconstructed in [4] by a method based on the $E_{7(+7)} \times SU(8)$ symmetry.

Field Content

The field content of this theory is a vierbein $e_\mu{}^a(x)$, 8 Majorana Rarita–Schwinger fields $\psi_\mu(x)$, 28 vector fields $B_\mu(x)$, 56 Majorana spinor fields $\lambda(x)$, 70 real scalar fields $\phi(x)$ as shown in Table 3.3. It is convenient to express the fermionic fields by Weyl spinors. We denote the Weyl Rarita–Schwinger fields obtained by multiplying the Majorana Rarita–Schwinger fields by the negative chirality projection operator $\frac{1}{2}(1 - \gamma_5)$ as $\psi_\mu{}^i(x)$ ($i = 1, 2, \ldots, 8$). $\psi_\mu{}^i(x)$ belong to the fundamental representation $\mathbf{8}$ of $SU(8)$. The charge conjugations $\psi_{\mu i}(x) = (\psi_\mu{}^i)^c(x)$ have positive chirality and belong to the representation $\bar{\mathbf{8}}$ of $SU(8)$. Similarly, the spinor fields are denoted by $\lambda^{ijk}(x)$ with negative chirality and their charge conjugations $\lambda_{ijk}(x) = (\lambda^{ijk})^c(x)$ with positive chirality. The indices ijk of these fields are totally antisymmetric and the spinor fields $\lambda^{ijk}(x)$ and $\lambda_{ijk}(x)$ belong to the representations $\mathbf{56}$ and $\overline{\mathbf{56}}$ of $SU(8)$, respectively. The 28 vector fields are $U(1)^{28}$ gauge fields and are denoted as $B_\mu{}^{IJ}(x) = -B_\mu{}^{JI}(x)$ ($I, J = 1, 2, \ldots, 8$). The field strengths of these fields $F_{\mu\nu}{}^{IJ} = 2\partial_{[\mu} B_{\nu]}^{IJ}$ and the tensors $G_{\mu\nu IJ}$ defined in (4.23) together belong to the representation $\mathbf{56}$ of $E_{7(+7)}$. The scalar fields are described by the $E_{7(+7)}/SU(8)$ non-linear sigma model as we will explain below, and transform by $\mathbf{56}$ under $E_{7(+7)}$ and by $\mathbf{28} + \overline{\mathbf{28}}$ under $SU(8)$. The gravitational field is invariant under both of $E_{7(+7)}$ and $SU(8)$.

$E_{7(+7)}/SU(8)$ Non-linear Sigma Model

The scalar fields are described by the $E_{7(+7)}/SU(8)$ non-linear sigma model. $E_{7(+7)}$ is a non-compact Lie group of dimension 133, and one of the real forms of the exceptional group E_7. The suffix $(+7)$ means that the difference of the numbers of positive and negative eigenvalues of the Killing form is 7. $E_{7(+7)}$ is a subgroup of $Sp(56, \mathbb{R})$, which is the largest possible duality group for 28 vector fields in $D = 4$ as we saw in Sect. 4.2. We can use the representation matrices of the Lie algebra of $Sp(56, \mathbb{R})$ with certain additional conditions imposed to represent those of $E_{7(+7)}$. In a complex basis as in (4.47) a general element of the $Sp(56, \mathbb{R})$ Lie algebra can be written as

$$X = \begin{pmatrix} \Lambda^{IJ}{}_{KL} & \Sigma^{IJKL} \\ \Sigma_{IJKL} & \Lambda_{IJ}{}^{KL} \end{pmatrix}, \quad \Lambda_{IJ}{}^{KL} = (\Lambda^{IJ}{}_{KL})^* = -\Lambda^{KL}{}_{IJ}, \quad \Sigma^{IJKL} = (\Sigma_{IJKL})^* = \Sigma^{KLIJ}, \tag{4.61}$$

where Λ and Σ are 28×28 complex matrices. The rows of the 56×56 matrix X are represented by the upper IJ and the lower IJ, and the columns by the lower KL and the upper KL, where $I, J, K, L = 1, 2, \ldots, 8$. These suffixes of Λ and Σ are antisymmetric in IJ and KL, e.g., $\Lambda^{IJ}{}_{KL} = -\Lambda^{JI}{}_{KL} = -\Lambda^{IJ}{}_{LK}$.[1] Elements of

---

[1] When Λ and Σ are components of a matrix like X in (4.61), the indices are restricted as $I < J$, $K < L$. Therefore, summations of indices for a product of two matrices are

the $E_{7(+7)}$ Lie algebra are given by (4.61) satisfying the additional conditions

$$\Lambda^{IJ}{}_{KL} = 4\delta^{[I}_{[K}\Lambda^{J]}{}_{L]}, \quad (\Lambda^I{}_J)^* = -\Lambda^J{}_I, \quad \Lambda^I{}_I = 0,$$

$$\Sigma_{IJKL} = \frac{1}{4!}\eta\varepsilon_{IJKLMNPQ}(\Sigma_{MNPQ})^*. \tag{4.63}$$

The parameter η takes a value $+1$ or -1, and specifies an embedding of $E_{7(+7)}$ in $Sp(56, \mathbb{R})$. $\Lambda^I{}_J$ corresponds to the $SU(8)$ subalgebra of $E_{7(+7)}$ and has 63 independent components. Σ_{IJKL} is the orthogonal complement of $SU(8)$ in $E_{7(+7)}$ and has 70 independent components. $\Lambda^I{}_J$ and Σ_{IJKL} together have 133 independent components, which is the dimension of $E_{7(+7)}$. The eigenvalues of the Killing form are negative for Λ and positive for Σ, and the difference of the numbers of them is 7. A group element of $E_{7(+7)}$ can be expressed as e^X (For more details of $E_{7(+7)}$, see [3, 4, 6]).

The scalar fields are described by the field $V(x)$ taking values in $E_{7(+7)}$. As in (4.47) it can be written as

$$V(x) = \begin{pmatrix} u^{IJ}{}_{ij}(x) & v^{IJij}(x) \\ v_{IJij}(x) & u_{IJ}{}^{ij}(x) \end{pmatrix}, \tag{4.64}$$

where IJ represent the rows and ij represent the columns of the matrix V. The matrices u, v satisfy the conditions of $Sp(56, \mathbb{R})$ in (4.47)

$$u_{IJ}{}^{ij} = (u^{IJ}{}_{ij})^*, \quad v^{IJij} = (v_{IJij})^*,$$

$$\frac{1}{2}\left(u_{IJ}{}^{ij}u^{IJ}{}_{kl} - v^{IJij}v_{IJkl}\right) = 2\delta^i_{[k}\delta^j_{l]}, \quad u^{IJ}{}_{ij}v_{IJkl} = v_{IJij}u^{IJ}{}_{kl} \tag{4.65}$$

and the conditions of $E_{7(+7)}$ corresponding to (4.63).

To construct the Lagrangian and the supertransformation laws we need derivatives of the scalar fields. Following (4.3) we define $P_\mu{}^{ijkl}$ and $Q_\mu{}^i{}_j$ by

$$V^{-1}\partial_\mu V = \begin{pmatrix} 4\delta^{[i}_{[k}Q_\mu{}^{j]}{}_{l]} & P_\mu{}^{ijkl} \\ P_{\mu ijkl} & 4\delta^{[k}_{[i}Q_{\mu j]}{}^{l]} \end{pmatrix},$$

$$P_{\mu ijkl} = \frac{1}{2}\left(u^{IJ}{}_{ij}\partial_\mu v_{IJkl} - v_{IJij}\partial_\mu u^{IJ}{}_{kl}\right) = (P_\mu{}^{ijkl})^* = P_{\mu klij},$$

$$Q_\mu{}^i{}_j = \frac{1}{12}\left(u_{IJ}{}^{ik}\partial_\mu u^{IJ}{}_{jk} - v^{IJik}\partial_\mu v_{IJjk}\right) = (Q_{\mu i}{}^j)^* = -Q_{\mu j}{}^i. \tag{4.66}$$

The matrix $V^{-1}\partial_\mu V$ is an element of the $E_{7(+7)}$ Lie algebra and has the from in (4.61), (4.63). $Q_\mu{}^i{}_j$ takes values in the $SU(8)$ subalgebra and transforms as an $SU(8)$

$$\sum_{K<L}\Lambda^{IJ}{}_{KL}\Sigma^{KLMN} = \frac{1}{2}\sum_{K,L}\Lambda^{IJ}{}_{KL}\Sigma^{KLMN} \equiv \frac{1}{2}\Lambda^{IJ}{}_{KL}\Sigma^{KLMN}. \tag{4.62}$$

The unit matrix with suffixes IJ, KL is $2\delta^I_{[K}\delta^J_{L]} = \delta^I_K\delta^J_L - \delta^I_L\delta^J_K$.

gauge field. $P_{\mu ijkl}$ takes values in the orthogonal complement of $SU(8)$ in $E_{7(+7)}$ and transforms covariantly under the local $SU(8)$. They also satisfy $Q_\mu{}^i{}_i = 0$, $P_\mu{}^{ijkl} = \frac{1}{4!}\eta\varepsilon^{ijklmnpq}P_{\mu mnpq}$.

Lagrangian

The Lagrangian of this theory is

$$\mathcal{L} = eR - \frac{1}{24}eP_{\mu ijkl}P^{\mu ijkl} - e\bar{\psi}_{\mu i}\gamma^{\mu\nu\rho}D_\nu\psi_\rho{}^i - \frac{1}{6}e\bar{\lambda}_{ijk}\gamma^\mu D_\mu\lambda^{ijk}$$
$$+ \frac{1}{6\sqrt{2}}e\left(\bar{\psi}_{\mu i}\gamma^\nu\gamma^\mu\lambda_{jkl}P_\nu{}^{ijkl} + \bar{\lambda}^{jkl}\gamma^\mu\gamma^\nu\psi_\mu{}^i P_{vijkl}\right) - \frac{1}{16}eF_{\mu\nu}{}^{IJ}K_{IJ,KL}F^{\mu\nu KL}$$
$$- \frac{1}{8}eF_{\mu\nu}{}^{IJ}\left(N_+{}^{IJ}{}_{ij}H_+{}^{\mu\nu ij} + N_{-IJ}{}^{ij}H_-{}^{\mu\nu}{}_{ij}\right) + \cdots, \tag{4.67}$$

where $\cdots$ represent terms quartic in the fermionic fields. The second term is the kinetic term of the scalar fields and has the form (4.5) using $P_{\mu ijkl}$ defined above. The third and fourth terms are the kinetic terms of the fermionic fields. The covariant derivatives contain the $SU(8)$ gauge fields $Q_\mu{}^i{}_j$ in addition to the spin connection $\omega_{\mu ab}$. There are no minimal couplings of the fermionic fields to the vector fields $B_\mu{}^{IJ}$. The fifth term represents couplings of the bilinears of the fermionic fields and the derivatives of the scalar fields similar to (4.7). Other two terms depend on the vector fields. $K_{IJ,KL}$ in the kinetic term of the vector fields is defined by first considering

$$K_{IJ,KL} = [(u^\dagger + v^\dagger)^{-1}(u^\dagger - v^\dagger)]_{IJ,KL} \tag{4.68}$$

and then by replacing the imaginary unit i in it by the Hodge dual operation $*$ as in (4.50). The last term is the Pauli term, which represents couplings of the bilinears of the fermionic fields and the field strengths of the vector fields. The quantities appearing there are defined by using the scalar fields and the fermionic fields as

$$N_+{}^{IJ}{}_{ij} = [(u^\dagger + v^\dagger)^{-1}]^{IJ}{}_{ij},$$
$$H_{+\mu\nu}{}^{ij} = \frac{1}{2}\bar{\psi}_\rho{}^i\gamma^{[\rho}\gamma_{\mu\nu}\gamma^{\sigma]}\psi_\sigma{}^j + \frac{1}{2\sqrt{2}}\bar{\psi}_{\rho k}\gamma_{\mu\nu}\gamma^\rho\lambda^{ijk} - \frac{1}{144}\eta\varepsilon^{ijklmnpq}\bar{\lambda}_{klm}\gamma_{\mu\nu}\lambda_{npq},$$
$$N_{-IJ}{}^{ij} = (N_+{}^{IJ}{}_{ij})^*, \qquad H_{-\mu vij} = (H_{+\mu\nu}{}^{ij})^*. \tag{4.69}$$

By the Weyl conditions on the fermionic fields, $H_\pm$ satisfy the duality property $*H_\pm = \pm iH_\pm$.

We can show that this Lagrangian has the form (4.41) and therefore the field equations are invariant under $E_{7(+7)}$. The first five terms of (4.67) are manifestly invariant under $E_{7(+7)}$ and correspond to $\mathcal{L}_{inv}$ in (4.41). The vector kinetic term in (4.67) is exactly the first term of (4.41) with K in (4.50). To compare the Pauli terms we define (H^{IJ}, I_{IJ}) by

$$\begin{pmatrix} H + iI \\ H - iI \end{pmatrix} = V\begin{pmatrix} iH_+ \\ -iH_- \end{pmatrix}. \tag{4.70}$$

This (H^{IJ}, I_{IJ}) is invariant under $SU(8)$ and transforms in the same way as (F^{IJ}, G_{IJ}) under $E_{7(+7)}$. Then, by the above duality property of $H_{\pm}$, the combination in (4.41) becomes

$$I_{IJ} - \frac{1}{2}K_{IJ,KL}*H^{KL} = -\frac{1}{2}\left(N_+{}^{IJ}{}_{ij}H_+{}^{ij} + N_{-IJ}^{ij}H_{-ij}\right), \tag{4.71}$$

which is the factor in the last term of (4.67). To derive (4.71) we have also used the fact that N_+ in (4.69) can be written as $N_+ = \frac{1}{2}\left[(u-v) + \bar{K}(u+v)\right]$, where $\bar{K}$ is (4.68) before replacing the imaginary unit i by the Hodge dual operation $*$. The $*H(I - K*H)$ term in (4.41) is quartic in the fermionic fields and is included in the terms $\cdots$ of (4.67).

Local Symmetries

The Lagrangian (4.67) is invariant up to total divergences under the general coordinate transformation δ_G, the local Lorentz transformation δ_L, the $U(1)^{28}$ gauge transformation δ_g, the local $SU(8)$ transformation $\delta_{SU(8)}$ and the local supertransformation δ_Q. The local supertransformation is given by

$$\delta_Q e_\mu{}^a = \frac{1}{4}\left(\bar{\varepsilon}_i\gamma^a\psi_\mu{}^i + \bar{\varepsilon}^i\gamma^a\psi_{\mu i}\right),$$

$$\delta_Q B_\mu{}^{IJ} = -\frac{1}{2}(u^{IJ}{}_{ij} + v_{IJij})\left(\bar{\varepsilon}^i\psi_\mu{}^j - \frac{1}{2\sqrt{2}}\bar{\varepsilon}_k\gamma_\mu\lambda^{ijk}\right) + \text{c.c.},$$

$$V^{-1}\delta_Q V = \begin{pmatrix} 0 & (W_{ijkl})^* \\ W_{ijkl} & 0 \end{pmatrix}, \quad W_{ijkl} = \sqrt{2}\left(\bar{\varepsilon}_{[i}\lambda_{jkl]} + \frac{1}{4!}\eta\varepsilon_{ijklmnpq}\bar{\varepsilon}^m\lambda^{npq}\right),$$

$$\delta_Q\psi_\mu{}^i = D_\mu\varepsilon^i + \frac{1}{16}\gamma^{\rho\sigma}\gamma_\mu F_{\rho\sigma}{}^{IJ}N_{-IJ}{}^{ij}\varepsilon_j + \cdots,$$

$$\delta_Q\lambda^{ijk} = -\frac{1}{\sqrt{2}}\gamma^\mu\varepsilon_l P_\mu{}^{ijkl} - \frac{3}{8\sqrt{2}}\gamma^{\mu\nu}F_{\mu\nu}{}^{IJ}N_{-IJ}{}^{[ij}\varepsilon^{k]} + \cdots, \tag{4.72}$$

where $\cdots$ denote terms quadratic in the fermionic fields. The transformation parameter $\varepsilon^i(x)$ is a Weyl spinor of negative chirality while its charge conjugation $(\varepsilon^i(x))^c = \varepsilon_i(x)$ is a Weyl spinor of positive chirality.

The local symmetry transformations form a closed algebra in commutation relations. In particular, the commutator of two local supertransformations is

$$[\delta_Q(\varepsilon_1), \delta_Q(\varepsilon_2)] = \delta_G(\xi) + \delta_L(\lambda) + \delta_g(\zeta) + \delta_{SU(8)}(\Lambda) + \delta_Q(\varepsilon) \tag{4.73}$$

when the field equations are used. The transformation parameters on the right-hand side are given by

$$\xi^\mu = \frac{1}{4}\left(\bar{\varepsilon}_{2i}\gamma^\mu \varepsilon_1^i + \bar{\varepsilon}_2^i \gamma^\mu \varepsilon_{1i}\right), \quad \Lambda^i{}_j = -\xi^\mu Q_\mu{}^i{}_j + \cdots, \quad \varepsilon^i = \cdots,$$

$$\lambda_{ab} = -\xi^\mu \omega_{\mu ab} - \frac{1}{64}\left(\bar{\varepsilon}_{2i}\gamma_{[a}\gamma^{\mu\nu}\gamma_{b]}\varepsilon_{1j}F_{\mu\nu}{}^{IJ}N_{-IJ}{}^{ij} + \text{c.c.}\right) + \cdots,$$

$$\zeta^{IJ} = -\xi^\mu B_\mu{}^{IJ} - \frac{1}{2}\left[(u^{IJ}{}_{ij} + v_{IJij})\bar{\varepsilon}_2^i \varepsilon_1^j + \text{c.c.}\right] + \cdots, \tag{4.74}$$

where $\cdots$ denote terms depending on the fermionic fields.

References

1. C.G. Callan Jr, S. Coleman, J. Wess, B. Zumino, Structure of phenomenological Lagrangians. 2. Phys. Rev. **177**, 2247 (1969)
2. S. Coleman, J. Wess, B. Zumino, Structure of phenomenological Lagrangians. 1. Phys. Rev. **177**, 2239 (1969)
3. E. Cremmer, B. Julia, The $SO(8)$ supergravity. Nucl. Phys. **B159**, 141 (1979)
4. B. de Wit, H. Nicolai, $\mathcal{N} = 8$ supergravity. Nucl. Phys. **B208**, 323 (1982)
5. M.K. Gaillard, B. Zumino, Duality rotations for interacting fields. Nucl. Phys. **B193**, 221 (1981)
6. R. Gilmore, *Lie Groups, Lie Algebras, and Some of Their Applications* (John Wiley & Sons, New York, 1974)
7. Y. Tanii, $\mathcal{N} = 8$ supergravity in six-dimensions. Phys. Lett. **B145**, 197 (1984)
8. Y. Tanii, Introduction to supergravities in diverse dimensions. Soryushiron Kenkyu, Kyoto **96**, 315 (1998). [hep-th/9802138]

Chapter 5
Poincaré Supergravities in Higher Dimensions

5.1 General Structure of Poincaré Supergravities

In four and higher dimensions there exist supergravities listed in Table 3.3. Their Lagrangians (or field equations) and local symmetry transformation laws were constructed (For detailed references, see [12]). These theories have Minkowski spacetime as a solution of the field equations and are called Poincaré supergravities.

General structure of Poincaré supergravity is as follows. Its field content is a gravitational field, Rarita–Schwinger fields, antisymmetric tensor fields, vector fields, spinor fields and scalar fields as shown in Table 3.3. It has local symmetries under the general coordinate transformation, the local Lorentz transformation, the gauge transformation of the vector and antisymmetric tensor fields and the local super-transformation. It also has a global symmetry G given in Table 4.1. When it contains scalar fields, G is a non-compact group and the scalar fields are described by the G/H non-linear sigma model, where H is a maximal compact subgroup of G. As we discussed in Sect. 4.1, the G/H non-linear sigma model can be formulated in such a way that the theory has the global G symmetry and the local H symmetry. The bosonic fields other than the gravitational field and the scalar fields transform only under the global G, while the fermionic fields transform only under the local H. The gravitational field does not transform under either of G or H, and the scalar fields transform under both of them. In some of the theories in even dimensions $D = 2n$, G acts on antisymmetric tensor fields of rank $n - 1$ as the duality transformation as discussed in Sect. 4.2. In this case G is a symmetry of the field equations but not of the action.

In the following we discuss $D = 11$ and $D = 10$ supergravities, which are low energy effective theories of M theory and superstring theories. In $D = 11$ there exists $\mathcal{N} = 1$ supergravity corresponding to M theory. In $D = 10$ there exist $\mathcal{N} = (1, 1)$, $(2, 0)$ and $(1, 0)$ supergravities corresponding to type IIA, IIB superstrings and type I superstring/heterotic strings. $D \leq 9$ supergravities can be obtained from $D = 11, 10$ theories by dimensional reductions and subsequent truncations as we will discuss in Chap. 6.

Y. Tanii, *Introduction to Supergravity*, SpringerBriefs in Mathematical Physics, DOI: 10.1007/978-4-431-54828-7_5, © The Author(s) 2014

5.2 $D = 11$, $\mathcal{N} = 1$ Poincaré Supergravity

Poincaré supergravity constructed in the highest spacetime dimension is $D = 11$, $\mathcal{N} = 1$ theory [5]. This theory is a low energy effective theory of M theory. The field content is a gravitational field $e_\mu{}^a(x)$, a real antisymmetric tensor field $B_{\mu\nu\rho}(x)$ and a Majorana Rarita–Schwinger field $\psi_\mu(x)$. Since it does not contain scalar fields, the Lagrangian and the local supertransformation laws have relatively simple forms. The Lagrangian is

$$
\mathcal{L} = eR(\omega) - \frac{1}{48}eF_{\mu\nu\rho\sigma}F^{\mu\nu\rho\sigma} - \frac{1}{144^2}\varepsilon^{\mu_1\ldots\mu_{11}}F_{\mu_1\ldots\mu_4}F_{\mu_5\ldots\mu_8}B_{\mu_9\mu_{10}\mu_{11}}
$$
$$
- \frac{1}{2}e\bar{\psi}_\mu\gamma^{\mu\nu\rho}D_\nu\left(\frac{\omega+\hat{\omega}}{2}\right)\psi_\rho + \frac{1}{384}e\bar{\psi}_{[\mu}\gamma^\mu\gamma^{\rho\sigma\tau\lambda}\gamma^\nu\psi_{\nu]}(F_{\rho\sigma\tau\lambda} + \hat{F}_{\rho\sigma\tau\lambda}) ,
$$

$$\tag{5.1}$$

where $R(\omega)$ and $D_\mu(\omega)$ are the scalar curvature and the covariant derivative defined by the spin connection $\omega_{\mu ab}$. We have also defined

$$
\hat{\omega}_{\mu ab} = \omega^{(0)}_{\mu ab} + \frac{1}{8}(\bar{\psi}_\mu\gamma_a\psi_b - \bar{\psi}_\mu\gamma_b\psi_a + \bar{\psi}_a\gamma_\mu\psi_b) ,
$$
$$
\omega_{\mu ab} = \hat{\omega}_{\mu ab} + \frac{1}{16}\bar{\psi}_\rho\gamma_{\mu ab}{}^{\rho\sigma}\psi_\sigma ,
$$
$$
F_{\mu\nu\rho\sigma} = 4\partial_{[\mu}B_{\nu\rho\sigma]} , \qquad \hat{F}_{\mu\nu\rho\sigma} = F_{\mu\nu\rho\sigma} - \frac{3}{2}\bar{\psi}_{[\mu}\gamma_{\nu\rho}\psi_{\sigma]} , \tag{5.2}
$$

where $\omega^{(0)}_{\mu ab}$ is the spin connection without torsion. If we regard the action S made from this Lagrangian as a functional of $e_\mu{}^a$, ψ_μ, $B_{\mu\nu\rho}$ and $\omega_{\mu ab}$, then it satisfies $\frac{\delta S}{\delta\omega_{\mu ab}} = 0$. The third term of (5.1) containing the Levi-Civita symbol is called the Chern–Simons term.

The Lagrangian (5.1) is invariant up to total divergences under the general coordinate transformation $\delta_G(\xi)$, the local Lorentz transformation $\delta_L(\lambda)$ and the gauge transformation of the antisymmetric tensor field $\delta_g(\zeta)$ with a parameter $\zeta_{\mu\nu}(x)$ given by

$$
\delta_g e_\mu{}^a = 0 , \quad \delta_g B_{\mu\nu\rho} = 3\,\partial_{[\mu}\zeta_{\nu\rho]} , \quad \delta_g\psi_\mu = 0 . \tag{5.3}
$$

It is also invariant under the local supertransformation $\delta_Q(\varepsilon)$ with a Majorana spinor parameter $\varepsilon(x)$

$$
\delta_Q e_\mu{}^a = \frac{1}{4}\bar{\varepsilon}\gamma^a\psi_\mu , \qquad \delta_Q B_{\mu\nu\rho} = \frac{3}{4}\bar{\varepsilon}\gamma_{[\mu\nu}\psi_{\rho]} ,
$$
$$
\delta_Q\psi_\mu = D_\mu(\hat{\omega})\varepsilon - \frac{1}{288}\left(\gamma_\mu{}^{\nu\rho\sigma\tau} - 8\delta_\mu^\nu\gamma^{\rho\sigma\tau}\right)\varepsilon\,\hat{F}_{\nu\rho\sigma\tau} . \tag{5.4}
$$

The commutator of two local supertransformations is

$$[\delta_Q(\varepsilon_1), \delta_Q(\varepsilon_2)] = \delta_G(\xi) + \delta_L(\lambda) + \delta_g(\zeta) + \delta_Q(\varepsilon) , \tag{5.5}$$

where the parameters on the right-hand side are

$$\xi^\mu = \frac{1}{4}\bar{\varepsilon}_2\gamma^\mu\varepsilon_1 , \quad \zeta_{\mu\nu} = -\xi^\rho B_{\rho\mu\nu} + \frac{1}{4}\bar{\varepsilon}_2\gamma_{\mu\nu}\varepsilon_1 , \quad \varepsilon = -\xi^\mu\psi_\mu ,$$

$$\lambda_{ab} = -\xi^\mu\hat{\omega}_{\mu ab} + \frac{1}{24}\bar{\varepsilon}_2\gamma^{\rho\sigma}\varepsilon_1\hat{F}_{ab\rho\sigma} + \frac{1}{576}\bar{\varepsilon}_2\gamma_{ab}{}^{\mu\nu\rho\sigma}\varepsilon_1\hat{F}_{\mu\nu\rho\sigma} . \tag{5.6}$$

To derive the commutation relation (5.5) we have to use the field equation of the Rarita–Schwinger field.

The field equations of this theory have an 11-dimensional Minkowski space-time solution $e_\mu{}^a = \delta_\mu^a$, $B_{\mu\nu\rho} = 0$, $\psi_\mu = 0$. The commutation relations of the local symmetry transformations in a background of this solution turn out to be the super Poincaré algebra as in $D = 4$, $\mathcal{N} = 1$ supergravity in Sect. 2.3. The gauge transformation $\delta_g(\zeta)$ with the parameter $\zeta_{\mu\nu} = \frac{1}{4}\bar{\varepsilon}_2\gamma_{\mu\nu}\varepsilon_1$ in (5.5) represents an anti-symmetric tensor central charge mentioned at the end of Sect. 3.2. Similar central charges also appear in supergravities in other dimensions.

5.3 $D = 10$, $\mathcal{N} = (1, 1)$ Poincaré Supergravity

$D = 10$, $\mathcal{N} = (1, 1)$ Poincaré supergravity [2, 6, 10] is a low energy effective theory of type IIA superstring theory. This theory is a vector-like theory, which is symmetric in positive and negative chiralities, and can be obtained from $D = 11$, $\mathcal{N} = 1$ supergravity by a dimensional reduction as we discuss in Sect. 6.3. The field content is a gravitational field $e_\mu{}^a(x)$, real antisymmetric tensor fields $B_{\mu\nu\rho}(x)$, $B_{\mu\nu}(x)$, a real vector field $B_\mu(x)$, a real scalar field $\phi(x)$, a Majorana Rarita–Schwinger field $\psi_\mu(x)$ and a Majorana spinor field $\lambda(x)$. A Majorana spinor can be also represented by two Majorana–Weyl spinors with positive and negative chiralities. The Lagrangian is

$$\mathcal{L} = eR - \frac{1}{2}e\partial_\mu\phi\partial^\mu\phi - \frac{1}{4}e\,e^{\frac{3}{2}\phi}F_{\mu\nu}F^{\mu\nu} - \frac{1}{12}e\,e^{-\phi}F_{\mu\nu\rho}F^{\mu\nu\rho}$$

$$- \frac{1}{48}e\,e^{\frac{1}{2}\phi}F_{\mu\nu\rho\sigma}F^{\mu\nu\rho\sigma} - \frac{1}{144}\varepsilon^{\mu_1\cdots\mu_{10}}\partial_{\mu_1}B_{\mu_2\mu_3\mu_4}\partial_{\mu_5}B_{\mu_6\mu_7\mu_8}B_{\mu_9\mu_{10}}$$

$$- \frac{1}{2}e\bar{\psi}_\mu\gamma^{\mu\nu\rho}D_\nu\psi_\rho - \frac{1}{2}e\bar{\lambda}\gamma^\mu D_\mu\lambda - \frac{1}{2\sqrt{2}}e\bar{\psi}_\mu\gamma^\nu\gamma^\mu\lambda\,\partial_\nu\phi$$

$$- \frac{1}{16}e\,e^{\frac{3}{4}\phi}\left(\bar{\psi}_{[\mu}\gamma^\mu\gamma^{\rho\sigma}\gamma^\nu\bar{\gamma}\psi_{\nu]} + \frac{3}{\sqrt{2}}\bar{\psi}_\mu\gamma^{\rho\sigma}\gamma^\mu\bar{\gamma}\lambda + \frac{5}{4}\bar{\lambda}\gamma^{\rho\sigma}\bar{\gamma}\lambda\right)F_{\rho\sigma}$$

$$+ \frac{1}{48}e\,e^{-\frac{1}{2}\phi}\left(\bar{\psi}_{[\mu}\gamma^\mu\gamma^{\rho\sigma\tau}\gamma^\nu\bar{\gamma}\psi_{\nu]} + \sqrt{2}\bar{\psi}_\mu\gamma^{\rho\sigma\tau}\gamma^\mu\bar{\gamma}\lambda\right)F_{\rho\sigma\tau}$$

$$+ \frac{1}{192}e\,e^{\frac{1}{4}\phi}\left(\bar{\psi}_{[\mu}\gamma^\mu\gamma^{\rho\sigma\tau\lambda}\gamma^\nu\psi_{\nu]} - \frac{1}{\sqrt{2}}\bar{\psi}_\mu\gamma^{\rho\sigma\tau\lambda}\gamma^\mu\lambda + \frac{3}{4}\bar{\lambda}\gamma^{\rho\sigma\tau\lambda}\lambda\right)F_{\rho\sigma\tau\lambda}$$

$$+ \cdots , \tag{5.7}$$

Table 5.1 $\mathbb{R}^+$ weights of $D = 10$, $\mathcal{N} = (1, 1)$ supergravity

Field	$e_\mu{}^a$	$B_{\mu\nu\rho}$	$B_{\mu\nu}$	B_μ	e^ϕ	ψ_μ	λ
w	0	$-\frac{1}{4}$	$\frac{1}{2}$	$-\frac{3}{4}$	1	0	0

where $\cdots$ denote terms higher order in the fermionic fields. The field strengths of the vector and antisymmetric tensor fields are

$$F_{\mu\nu} = 2\partial_{[\mu}B_{\nu]} \,, \quad F_{\mu\nu\rho} = 3\partial_{[\mu}B_{\nu\rho]} \,, \quad F_{\mu\nu\rho\sigma} = 4\partial_{[\mu}B_{\nu\rho\sigma]} + 4B_{[\mu}F_{\nu\rho\sigma]} \,. \quad (5.8)$$

The covariant derivatives D_μ on the fermionic fields contain only the spin connection as a gauge field. $\bar{\gamma}$ is the chirality matrix in (3.11).

The Lagrangian (5.7) is invariant under the global $\mathbb{R}^+$ transformation $\Phi(x) \to e^{w\Lambda}\Phi(x)$, where the weight w of each field $\Phi(x)$ is given in Table 5.1. It is also invariant up to total divergences under the general coordinate transformation $\delta_G(\xi)$, the local Lorentz transformation $\delta_L(\lambda)$, the gauge transformation of the vector and antisymmetric tensor fields $\delta_g(\zeta)$ and the local supertransformation $\delta_Q(\varepsilon)$. The gauge transformation has three kinds of transformation parameters $\zeta(x)$, $\zeta_\mu(x)$, $\zeta_{\mu\nu}(x)$ and is given by

$$\delta_g B_\mu = \partial_\mu\zeta \,, \quad \delta_g B_{\mu\nu} = 2\partial_{[\mu}\zeta_{\nu]} \,, \quad \delta_g B_{\mu\nu\rho} = 3\partial_{[\mu}\zeta_{\nu\rho]} + 3B_{[\mu\nu}\partial_{\rho]}\zeta \,. \quad (5.9)$$

The field strengths (5.8) are invariant under this gauge transformation. The second term of $\delta_g B_{\mu\nu\rho}$ is needed because of the term $4B_{[\mu}F_{\nu\rho\sigma]}$ in $F_{\mu\nu\rho\sigma}$. The local supertransformation with a Majorana spinor parameter $\varepsilon(x)$ is given by

$$\delta_Q e_\mu{}^a = \frac{1}{4}\bar{\varepsilon}\gamma^a\psi_\mu \,, \quad \delta_Q\phi = -\frac{1}{2\sqrt{2}}\bar{\varepsilon}\lambda \,,$$

$$\delta_Q B_\mu = -\frac{1}{4}e^{-\frac{3}{4}\phi}\bar{\varepsilon}\bar{\gamma}\psi_\mu + \frac{3}{8\sqrt{2}}e^{-\frac{3}{4}\phi}\bar{\varepsilon}\gamma_\mu\bar{\gamma}\lambda \,,$$

$$\delta_Q B_{\mu\nu} = -\frac{1}{2}e^{\frac{1}{2}\phi}\bar{\varepsilon}\bar{\gamma}\gamma_{[\mu}\psi_{\nu]} + \frac{1}{4\sqrt{2}}e^{\frac{1}{2}\phi}\bar{\varepsilon}\gamma_{\mu\nu}\bar{\gamma}\lambda \,,$$

$$\delta_Q B_{\mu\nu\rho} = \frac{3}{4}e^{-\frac{1}{4}\phi}\bar{\varepsilon}\gamma_{[\mu\nu}\psi_{\rho]} + \frac{1}{8\sqrt{2}}e^{-\frac{1}{4}\phi}\bar{\varepsilon}\gamma_{\mu\nu\rho}\lambda + 3B_{[\mu}\delta_Q B_{\nu\rho]} \,,$$

$$\delta_Q\psi_\mu = D_\mu\varepsilon + \frac{1}{64}e^{\frac{3}{4}\phi}\left(\gamma_\mu{}^{\nu\rho} - 14\delta_\mu^\nu\gamma^\rho\right)\bar{\gamma}\varepsilon F_{\nu\rho}$$

$$+ \frac{1}{96}e^{-\frac{1}{2}\phi}\left(\gamma_\mu{}^{\nu\rho\sigma} - 9\delta_\mu^\nu\gamma^{\rho\sigma}\right)\bar{\gamma}\varepsilon F_{\nu\rho\sigma}$$

$$- \frac{1}{768}e^{\frac{1}{4}\phi}\left(3\gamma_\mu{}^{\nu\rho\sigma\tau} - 20\delta_\mu^\nu\gamma^{\rho\sigma\tau}\right)\varepsilon F_{\nu\rho\sigma\tau} + \cdots \,,$$

$$\delta_Q\lambda = -\frac{1}{2\sqrt{2}}\gamma^\mu\varepsilon\partial_\mu\phi - \frac{3}{16\sqrt{2}}e^{\frac{3}{4}\phi}\gamma^{\mu\nu}\bar{\gamma}\varepsilon F_{\mu\nu} + \frac{1}{24\sqrt{2}}e^{-\frac{1}{2}\phi}\gamma^{\mu\nu\rho}\bar{\gamma}\varepsilon F_{\mu\nu\rho}$$

$$+ \frac{1}{192\sqrt{2}} e^{\frac{1}{4}\phi} \gamma^{\mu\nu\rho\sigma} \varepsilon F_{\mu\nu\rho\sigma} + \cdots , \tag{5.10}$$

where $\cdots$ denote terms higher order in the fermionic fields. The commutator of two supertransformations is

$$[\delta_Q(\varepsilon_1), \delta_Q(\varepsilon_2)] = \delta_G(\xi) + \delta_L(\lambda) + \delta_g(\zeta) + \delta_Q(\varepsilon) . \tag{5.11}$$

The transformation parameters on the right-hand side are

$$\xi^\mu = \frac{1}{4} \bar{\varepsilon}_2 \gamma^\mu \varepsilon_1 , \qquad \varepsilon = \cdots ,$$

$$\lambda_{ab} = - \xi^\mu \omega_{\mu ab} - \frac{1}{128} e^{\frac{3}{4}\phi} \bar{\varepsilon}_2 \left(\gamma_{ab}{}^{\mu\nu} + 14 e_a{}^\mu e_b{}^\nu \right) \bar{\gamma} \varepsilon_1 F_{\mu\nu} ,$$

$$- \frac{1}{192} e^{-\frac{1}{2}\phi} \bar{\varepsilon}_2 \left(\gamma_{ab}{}^{\mu\nu\rho} + 18 e_a{}^\mu e_b{}^\nu \gamma^\rho \right) \bar{\gamma} \varepsilon_1 F_{\mu\nu\rho}$$

$$+ \frac{1}{512} e^{\frac{1}{4}\phi} \bar{\varepsilon}_2 \left(\gamma_{ab}{}^{\mu\nu\rho\sigma} + 20 e_a{}^\mu e_b{}^\nu \gamma^{\rho\sigma} \right) \varepsilon_1 F_{\mu\nu\rho\sigma} + \cdots ,$$

$$\zeta = - \frac{1}{4} e^{-\frac{3}{4}\phi} \bar{\varepsilon}_2 \bar{\gamma} \varepsilon_1 - \xi^\mu B_\mu + \cdots , \qquad \zeta_\mu = - \frac{1}{4} e^{\frac{1}{2}\phi} \bar{\varepsilon}_2 \gamma_\mu \bar{\gamma} \varepsilon_1 - \xi^\nu B_{\nu\mu} + \cdots ,$$

$$\zeta_{\mu\nu} = \frac{1}{4} e^{-\frac{1}{4}\phi} \bar{\varepsilon}_2 \gamma_{\mu\nu} \varepsilon_1 - \xi^\rho B_{\rho\mu\nu} - \zeta B_{\mu\nu} - \frac{1}{2} e^{\frac{1}{2}\phi} \bar{\varepsilon}_2 \gamma_{[\mu} \bar{\gamma} \varepsilon_1 B_{\nu]} + \cdots , \tag{5.12}$$

where $\cdots$ denote terms depending on the fermionic fields. To derive the commutation relation (5.11) we have to use the field equations of the fermionic fields.

String Frame

$D = 10$, $\mathscr{N} = (1, 1)$ supergravity is a low energy effective theory of type IIA superstring theory. In string theory one often uses a Lagrangian different from (5.7), whose bosonic terms are

$$\mathscr{L} = e\, e^{-2\phi} \left[R + 4 \partial_\mu \phi \partial^\mu \phi - \frac{1}{12} F_{\mu\nu\rho} F^{\mu\nu\rho} \right] - \frac{1}{48} e F_{\mu\nu\rho\sigma} F^{\mu\nu\rho\sigma} - \frac{1}{4} e F_{\mu\nu} F^{\mu\nu}$$

$$- \frac{1}{144} \varepsilon^{\mu_1 \cdots \mu_{10}} \partial_{\mu_1} B_{\mu_2\mu_3\mu_4} \partial_{\mu_5} B_{\mu_6\mu_7\mu_8} B_{\mu_9\mu_{10}} + \cdots . \tag{5.13}$$

This Lagrangian has the scalar field factor $e^{-2\phi}$ in the curvature term. The terms depending only on the fields $e_\mu{}^a$, ϕ, $B_{\mu\nu}$ in the NS–NS sector of superstring have the factor $e^{-2\phi}$, but those depending also on the fields B_μ, $B_{\mu\nu\rho}$ in the R–R sector do not have such a factor. In superstring theory the scalar field ϕ is called the dilaton field and a background value of e^ϕ represents a coupling constant of string interactions.

The Lagrangians (5.7) and (5.13) are related by a field redefinition of the gravitational field. Making the Weyl transformation

$$e_\mu{}^a \to e^{-\frac{1}{4}\phi} e_\mu{}^a \tag{5.14}$$

in (5.7) and using (1.26) we obtain (5.13). The metric in the Lagrangian (5.7) with the standard Einstein term is called the Einstein frame while the metric in (5.13) is called the string frame.

5.4 $D = 10$, $\mathcal{N} = (2, 0)$ Poincaré Supergravity

$D = 10$, $\mathcal{N} = (2, 0)$ Poincaré supergravity [8, 13] is a low energy effective theory of type IIB superstring theory. This theory is a chiral theory, which is asymmetric in positive and negative chiralities. The field content is a gravitational field $e_\mu{}^a(x)$, a real antisymmetric tensor field $B_{\mu\nu\rho\sigma}(x)$, a complex antisymmetric tensor field $B_{\mu\nu}(x)$, a complex scalar field $\tau(x)$, a Weyl Rarita–Schwinger field of positive chirality $\psi_\mu(x)$ and a Weyl spinor field of negative chirality $\lambda(x)$. A Weyl spinor can be also represented by two Majorana–Weyl spinors with the same chirality. The field strength of the fourth rank antisymmetric tensor field $B_{\mu\nu\rho\sigma}(x)$ satisfies a self-duality equation. Because of this condition a simple covariant Lagrangian of this theory is not known (See Sect. 1.4). However, the covariant field equations were explicitly constructed. The field equations are invariant under the global $SU(1, 1) \sim SL(2, \mathbb{R})$ transformation in addition to the local transformations.

$SU(1, 1)/U(1)$ Non-linear Sigma Model

The scalar field is described by the $SU(1, 1)/U(1) \sim SL(2, \mathbb{R})/SO(2)$ non-linear sigma model. As in Sect. 4.1 the scalar field can be represented by an $SU(1, 1)$ matrix-valued field

$$V(x) = \begin{pmatrix} V^1{}_-(x) & V^1{}_+(x) \\ V^2{}_-(x) & V^2{}_+(x) \end{pmatrix} , \tag{5.15}$$

where the components $V^\alpha{}_\pm(x)$ ($\alpha = 1, 2$) are complex scalar fields. The condition that $V(x)$ is an $SU(1, 1)$ matrix [(A.7) in Appendix A] is

$$\varepsilon_{\alpha\beta} V^\alpha{}_- V^\beta{}_+ = 1 , \qquad (V^1{}_\pm)^* = V^2{}_\mp , \tag{5.16}$$

where $\varepsilon_{\alpha\beta} = -\varepsilon_{\beta\alpha}, \varepsilon_{12} = +1$. The matrix-valued scalar field $V(x)$ transforms under the global $G = SU(1, 1)$ transformation g and the local $H = U(1)$ transformation $h(x)$ as in (4.1). Explicit forms of the matrices g and $h(x)$ are

$$g = \begin{pmatrix} u & v \\ v^* & u^* \end{pmatrix} \quad \left(|u|^2 - |v|^2 = 1 \right) , \quad h(x) = \begin{pmatrix} e^{i\Sigma(x)} & 0 \\ 0 & e^{-i\Sigma(x)} \end{pmatrix} , \tag{5.17}$$

where u, v are complex numbers and $\Sigma(x)$ is a real function.

Following (4.3) we decompose $V^{-1} \partial_\mu V$, which belongs to the $SU(1, 1)$ Lie algebra, into a part in the $U(1)$ Lie algebra and a part in its orthogonal complement as

Table 5.2 $SU(1, 1)$ representations and $U(1)$ charges of $D = 10$, $\mathcal{N} = (2, 0)$ supergravity

Field	$e_\mu{}^a$	$B_{\mu\nu\rho\sigma}$	$B^\alpha_{\mu\nu}$	$V^\alpha{}_\pm$	ψ_μ	λ
$SU(1, 1)$	1	1	2	2	1	1
$U(1)$	0	0	0	± 1	$\frac{1}{2}$	$\frac{3}{2}$

$$V^{-1}\partial_\mu V = \begin{pmatrix} -iQ_\mu & 0 \\ 0 & iQ_\mu \end{pmatrix} + \begin{pmatrix} 0 & P_\mu \\ P^*_\mu & 0 \end{pmatrix} , \qquad (5.18)$$

where

$$P_\mu = -\varepsilon_{\alpha\beta} V^\alpha{}_+ \partial_\mu V^\beta{}_+ , \qquad Q_\mu = -i\varepsilon_{\alpha\beta} V^\alpha{}_- \partial_\mu V^\beta{}_+ . \qquad (5.19)$$

P_μ and Q_μ are invariant under the global $SU(1, 1)$ and transform under the local $U(1)$ as

$$P_\mu \to e^{2i\Sigma} P_\mu , \qquad Q_\mu \to Q_\mu + \partial_\mu \Sigma . \qquad (5.20)$$

We see that P_μ has $U(1)$ charge $+2$ and Q_μ transforms as a $U(1)$ gauge field. Q_μ can be used to define the covariant derivative. The covariant derivative on a field with $U(1)$ charge q is given by

$$D_\mu \Phi = \left(\partial_\mu - iq Q_\mu\right) \Phi . \qquad (5.21)$$

$SU(1, 1)$ representations and $U(1)$ charges of the fields are summarized in Table 5.2. The bosonic fields other than the gravitational field and the scalar fields are invariant under $U(1)$ and transform only under $SU(1, 1)$. The second rank anti-symmetric tensor fields $B^\alpha_{\mu\nu}(x)$ ($\alpha = 1, 2$) defined by $B^1_{\mu\nu} = B_{\mu\nu}$, $B^2_{\mu\nu} = B^*_{\mu\nu}$ transform as **2** of $SU(1, 1)$. The fermionic fields are invariant under $SU(1, 1)$ and transform only under $U(1)$. The scalar fields transform under both of $SU(1, 1)$ and $U(1)$. The gravitational field does not transform under either of $SU(1, 1)$ or $U(1)$.

Local Symmetries

The bosonic local symmetries are those under the general coordinate transformation $\delta_G(\xi)$, the local Lorentz transformation $\delta_L(\lambda)$, the gauge transformation of the anti-symmetric tensor fields $\delta_g(\zeta)$ and the $U(1)$ transformation of the $SU(1, 1)/U(1)$ non-linear sigma model $\delta_{U(1)}(\Sigma)$. The gauge transformation of the antisymmetric tensor fields is given by

$$\delta_g B^\alpha_{\mu\nu} = 2\partial_{[\mu}\zeta^\alpha_{\nu]} , \qquad \delta_g B_{\mu\nu\rho\sigma} = 4\partial_{[\mu}\zeta_{\nu\rho\sigma]} + 2i\varepsilon_{\alpha\beta}\zeta^\alpha_{[\mu} F^\beta_{\nu\rho\sigma]} , \qquad (5.22)$$

where $\zeta^\alpha_\mu(x)$ and $\zeta_{\mu\nu\rho}(x)$ are transformation parameters. Other fields are invariant under $\delta_g(\zeta)$. The field strengths of the antisymmetric tensor fields defined by

$$F^\alpha_{\mu\nu\rho} = 3\partial_{[\mu} B^\alpha_{\nu\rho]} , \qquad F_{\mu\nu\rho\sigma\tau} = 5\partial_{[\mu} B_{\nu\rho\sigma\tau]} - 5i\varepsilon_{\alpha\beta} B^\alpha_{[\mu\nu} F^\beta_{\rho\sigma\tau]} \qquad (5.23)$$

are invariant under (5.22). It is convenient to define

$$F_{\mu\nu\rho} = -\varepsilon_{\alpha\beta} V^{\alpha}{}_{+} F^{\beta}_{\mu\nu\rho} \,, \qquad (5.24)$$

which is $SU(1, 1)$ invariant and has $U(1)$ charge $+1$. By the condition (5.16) we find

$$F^{\alpha}_{\mu\nu\rho} = V^{\alpha}{}_{-} F_{\mu\nu\rho} + V^{\alpha}{}_{+} F^{*}_{\mu\nu\rho} \,. \qquad (5.25)$$

The $\mathcal{N} = (2, 0)$ local supertransformation $\delta_Q(\varepsilon)$ is given by

$$\delta_Q e_{\mu}{}^{a} = \frac{1}{4} \left(\bar{\varepsilon}\gamma^{a}\psi_{\mu} - \bar{\psi}_{\mu}\gamma^{a}\varepsilon \right) \,, \qquad \delta_Q V^{\alpha}{}_{+} = \frac{1}{2\sqrt{2}} V^{\alpha}{}_{-}\bar{\varepsilon}^{c}\lambda \,,$$

$$\delta_Q B^{\alpha}_{\mu\nu} = \frac{1}{4} \left(V^{\alpha}{}_{-}\bar{\varepsilon}\gamma_{\mu\nu}\lambda - V^{\alpha}{}_{+}\bar{\lambda}\gamma_{\mu\nu}\varepsilon \right)$$

$$+ \frac{1}{\sqrt{2}} \left(V^{\alpha}{}_{-}\bar{\varepsilon}^{c}\gamma_{[\mu}\psi_{\nu]} - V^{\alpha}{}_{+}\bar{\psi}_{[\nu}\gamma_{\mu]}\varepsilon^{c} \right) \,,$$

$$\delta_Q B_{\mu\nu\rho\sigma} = i\bar{\varepsilon}\gamma_{[\mu\nu\rho}\psi_{\sigma]} - i\bar{\psi}_{[\sigma}\gamma_{\mu\nu\rho]}\varepsilon + 3i\varepsilon_{\alpha\beta} B^{\alpha}_{[\mu\nu}\delta_Q B^{\beta}_{\rho\sigma]} \,,$$

$$\delta_Q \psi_{\mu} = D_{\mu}\varepsilon - \frac{1}{16 \times 5!} i\gamma^{\nu_1 \dots \nu_5}\gamma_{\mu}\varepsilon F_{\nu_1 \dots \nu_5}$$

$$+ \frac{1}{48\sqrt{2}} \left(\gamma_{\mu}{}^{\nu\rho\sigma} - 9\delta^{\nu}_{\mu}\gamma^{\rho\sigma} \right) \varepsilon^{c} F_{\nu\rho\sigma} + \cdots \,,$$

$$\delta_Q \lambda = \frac{1}{\sqrt{2}} \gamma^{\mu}\varepsilon^{c} P_{\mu} - \frac{1}{24} \gamma^{\mu\nu\rho}\varepsilon F_{\mu\nu\rho} + \cdots \,, \qquad (5.26)$$

where $\cdots$ denote terms higher order in the fermionic fields. The transformation parameter $\varepsilon(x)$ is a Weyl spinor of positive chirality and ε^{c} is its charge conjugation. The commutator of two supertransformations is

$$[\delta_Q(\varepsilon_1), \delta_Q(\varepsilon_2)] = \delta_G(\xi) + \delta_L(\lambda) + \delta_g(\zeta) + \delta_{U(1)}(\Sigma) + \delta_Q(\varepsilon) \,. \qquad (5.27)$$

The transformation parameters on the right-hand side are

$$\xi^{\mu} = \frac{1}{4} \left(\bar{\varepsilon}_2\gamma^{\mu}\varepsilon_1 - \bar{\varepsilon}_1\gamma^{\mu}\varepsilon_2 \right) \,,$$

$$\lambda_{ab} = -\xi^{\mu}\omega_{\mu ab} - \frac{1}{96} i \left(\bar{\varepsilon}_2\gamma^{\nu\rho\sigma}\varepsilon_1 - \bar{\varepsilon}_1\gamma^{\nu\rho\sigma}\varepsilon_2 \right) F_{\nu\rho\sigma ab}$$

$$- \frac{1}{96\sqrt{2}} \bar{\varepsilon}_2 \left(\gamma^{\nu\rho\sigma}{}_{ab} + 18e^{\rho}{}_{a}e^{\sigma}{}_{b}\gamma^{\nu} \right) \varepsilon^{c}_1 F_{\nu\rho\sigma}$$

$$+ \frac{1}{96\sqrt{2}} \bar{\varepsilon}^{c}_1 \left(\gamma^{\nu\rho\sigma}{}_{ab} + 18e^{\rho}{}_{a}e^{\sigma}{}_{b}\gamma^{\nu} \right) \varepsilon_2 F^{*}_{\nu\rho\sigma} + \cdots \,,$$

$$\zeta^{\alpha}_{\mu} = -\xi^{\nu} B^{\alpha}_{\nu\mu} - \frac{1}{2\sqrt{2}} \left(\bar{\varepsilon}^{c}_2\gamma_{\mu}\varepsilon_1 V^{\alpha}{}_{-} + \bar{\varepsilon}_2\gamma_{\mu}\varepsilon^{c}_1 V^{\alpha}{}_{+} \right) + \cdots \,,$$

$$\zeta_{\mu\nu\rho} = -\xi^\sigma B_{\sigma\mu\nu\rho} - \frac{1}{4}\mathrm{i}\left(\bar{\varepsilon}_2\gamma_{\mu\nu\rho}\varepsilon_1 - \bar{\varepsilon}_1\gamma_{\mu\nu\rho}\varepsilon_2\right)$$

$$- \frac{3}{4\sqrt{2}}\mathrm{i}\varepsilon_{\alpha\beta}B^\alpha_{[\mu\nu}\left(\bar{\varepsilon}^c_2\gamma_{\rho]}\varepsilon_1 V^\beta{}_- + \bar{\varepsilon}_2\gamma_{\rho]}\varepsilon^c_1 V^\beta{}_+\right) + \cdots ,$$

$$\Sigma = -\xi^\mu Q_\mu + \cdots , \quad \varepsilon = \ldots , \tag{5.28}$$

where $\cdots$ denote terms depending on the fermionic fields. To derive the commutation relation (5.27) we have to use the field equations of the fermionic fields and the self-duality equation of $B_{\mu\nu\rho\sigma}$, which we will discuss below.

Field Equations

The antisymmetric tensor field $B_{\mu\nu\rho\sigma}(x)$ satisfies a self-duality equation

$$\hat{F}_{\mu\nu\rho\sigma\tau} = *\hat{F}_{\mu\nu\rho\sigma\tau} , \tag{5.29}$$

where $*$ is the Hodge dual defined in (1.43), and $\hat{F}_{\mu\nu\rho\sigma\tau}$ is given by

$$\hat{F}_{\mu\nu\rho\sigma\tau} = F_{\mu\nu\rho\sigma\tau} - 5\mathrm{i}\bar{\psi}_{[\mu}\gamma_{\nu\rho\sigma}\psi_{\tau]} + \frac{1}{8}\mathrm{i}\bar{\lambda}\gamma_{\mu\nu\rho\sigma\tau}\lambda \tag{5.30}$$

in terms of the field strength $F_{\mu\nu\rho\sigma\tau}$ in (5.23). The self-duality equation (5.29) can be regarded as the field equation of $B_{\mu\nu\rho\sigma}(x)$. Applying a covariant derivative D_μ on (5.29) we obtain a second order Maxwell type equation

$$D_\tau F^{\tau\mu\nu\rho\sigma} = -\frac{1}{72}\mathrm{i}e^{-1}\varepsilon^{\mu\nu\rho\sigma\lambda_1\ldots\lambda_6}\varepsilon_{\alpha\beta}F^\alpha_{\lambda_1\lambda_2\lambda_3}F^\beta_{\lambda_4\lambda_5\lambda_6} + \cdots , \tag{5.31}$$

where $\cdots$ denote terms depending on the fermionic fields.

The field equations of other fields are

$$R_{\mu\nu} - \frac{1}{2}g_{\mu\nu}R = P^*_\mu P_\nu + P^*_\nu P_\mu - g_{\mu\nu}P^*_\rho P^\rho + \frac{1}{96}F_{\mu\rho\sigma\tau\lambda}F_\nu{}^{\rho\sigma\tau\lambda}$$

$$+ \frac{1}{4}\left(F^*_{\mu\rho\sigma}F_\nu{}^{\rho\sigma} + F^*_{\nu\rho\sigma}F_\mu{}^{\rho\sigma} - \frac{1}{3}g_{\mu\nu}F^*_{\rho\sigma\tau}F^{\rho\sigma\tau}\right) + \cdots ,$$

$$D_\rho F^{\rho\mu\nu} = P_\rho F^{*\rho\mu\nu} + \frac{1}{6}\mathrm{i}F^{\mu\nu\rho\sigma\tau}F_{\rho\sigma\tau} + \cdots ,$$

$$D_\mu P^\mu = -\frac{1}{12}F_{\mu\nu\rho}F^{\mu\nu\rho} + \cdots ,$$

$$\gamma^{\mu\nu\rho}\hat{D}_\nu\psi_\rho = \frac{1}{\sqrt{2}}\gamma^\nu\gamma^\mu\lambda^c P_\nu + \frac{1}{24}\gamma^{\nu\rho\sigma}\gamma^\mu\lambda F^*_{\nu\rho\sigma} + \cdots ,$$

$$\gamma^\mu\hat{D}_\mu\lambda = -\frac{1}{8\times 5!}\mathrm{i}\gamma^{\mu_1\ldots\mu_5}\lambda F_{\mu_1\ldots\mu_5} + \cdots . \tag{5.32}$$

The supercovariant derivatives $\hat{D}_\mu$ in the fermionic field equations are defined by

$$\hat{D}_{[\nu}\psi_{\rho]} = D_{[\nu}\psi_{\rho]} - \frac{1}{16 \times 5!}\, i\,\gamma^{\lambda_1\ldots\lambda_5}\gamma_{[\nu}\psi_{\rho]}F_{\lambda_1\ldots\lambda_5}$$

$$+ \frac{1}{48\sqrt{2}}\left(\gamma_\nu{}^{\lambda_1\lambda_2\lambda_3} - 9\delta^{\lambda_1}_{[\nu}\gamma^{\lambda_2\lambda_3]}\right)\psi^c_{\rho]}F_{\lambda_1\lambda_2\lambda_3} + \cdots,$$

$$\hat{D}_\mu\lambda = D_\mu\lambda - \frac{1}{\sqrt{2}}\gamma^\nu\psi^c_\mu P_\nu + \frac{1}{24}\gamma^{\nu\rho\sigma}\psi_\mu F_{\nu\rho\sigma} + \cdots. \tag{5.33}$$

The field equations (5.29), (5.32) are invariant under the global $SU(1,1)$ and the local transformations discussed above.

Lagrangian

Because of the self-duality of the antisymmetric tensor field $B_{\mu\nu\rho\sigma}(x)$ a simple covariant Lagrangian, from which all the field equations can be derived, is not known. However, a covariant Lagrangian which gives the field equations other than that of $B_{\mu\nu\rho\sigma}(x)$ is known. Such a Lagrangian is

$$\mathscr{L} = eR - 2eP^*_\mu P^\mu - \frac{1}{6}eF^*_{\mu\nu\rho}F^{\mu\nu\rho} - \frac{1}{4\times 5!}eF_{\mu\nu\rho\sigma\tau}F^{\mu\nu\rho\sigma\tau}$$

$$+ \frac{1}{144\times 4!}i\varepsilon^{\mu_1\ldots\mu_{10}}B_{\mu_1\ldots\mu_4}\varepsilon_{\alpha\beta}F^\alpha_{\mu_5\mu_6\mu_7}F^\beta_{\mu_8\mu_9\mu_{10}} - e\bar{\psi}_\mu\gamma^{\mu\nu\rho}D_\nu\psi_\rho$$

$$- e\bar{\lambda}\gamma^\mu D_\mu\lambda + \frac{1}{\sqrt{2}}e\bar{\psi}_\mu\gamma^\nu\gamma^\mu\lambda^c P_\nu + \frac{1}{\sqrt{2}}e\bar{\lambda}^c\gamma^\mu\gamma^\nu\psi_\mu P^*_\nu$$

$$+ \frac{1}{24\sqrt{2}}e\left(\bar{\psi}_{[\mu}\gamma^\mu\gamma^{\rho\sigma\tau}\gamma^\nu\psi^c_{\nu]} - \sqrt{2}\bar{\lambda}\gamma^\mu\gamma^{\rho\sigma\tau}\psi_\mu\right)F_{\rho\sigma\tau}$$

$$+ \frac{1}{24\sqrt{2}}e\left(\bar{\psi}^c_{[\mu}\gamma^\mu\gamma^{\rho\sigma\tau}\gamma^\nu\psi_{\nu]} + \sqrt{2}\bar{\psi}_\mu\gamma^{\rho\sigma\tau}\gamma^\mu\lambda\right)F^*_{\rho\sigma\tau}$$

$$+ \frac{1}{8\times 5!}ie\left(\bar{\psi}_{[\mu}\gamma^\mu\gamma^{\rho_1\ldots\rho_5}\gamma^\nu\psi_{\nu]} - \bar{\lambda}\gamma^{\rho_1\ldots\rho_5}\lambda\right)F_{\rho_1\ldots\rho_5} + \cdots, \tag{5.34}$$

where $\cdots$ denote terms higher order in the fermionic fields. The variations of this Lagrangian with respect to the fields other than $B_{\mu\nu\rho\sigma}(x)$ give the field equations in (5.32). The variation with respect to $B_{\mu\nu\rho\sigma}(x)$ gives the second order equation (5.31), which is consistent with the self-duality equation (5.29). Therefore, the variational equations derived from the Lagrangian (5.34) supplemented with the self-duality equation (5.29) give all the field equations.

Gauge Fixing of the Scalar Fields

We can fix a gauge for the local $U(1)$ transformation and describe the scalar fields by only physical degrees of freedom. In general, the $SU(1,1)$ matrix in (5.15) can be parametrized by a complex scalar field $B(x)$ and a real scalar field $\theta(x)$ as

$$V = \frac{1}{\sqrt{1 - |B|^2}} \begin{pmatrix} e^{-i\theta} & B e^{i\theta} \\ B^* e^{-i\theta} & e^{i\theta} \end{pmatrix} . \tag{5.35}$$

$B(x)$ takes values inside of a unit disc $|B| < 1$. It is easy to see that this V satisfies the condition (5.16). These fields transform under the local $U(1)$ transformation in (5.17) as $B \to B$, $\theta \to \theta + \Sigma$. Therefore, we can fix a gauge by imposing a certain condition on θ. We use the gauge fixing condition

$$e^{i\theta} = \left(\frac{1 + B^*}{1 + B} \right)^{\frac{1}{2}} . \tag{5.36}$$

After the gauge fixing we have to add compensating $U(1)$ transformations to the $SU(1, 1)$ transformation and the local supertransformation in order to preserve the gauge. Let us first consider the $SU(1, 1)$ transformation. An infinitesimal $SU(1, 1)$ transformation is given by the matrix g in (5.17) with $u = 1 + i\beta$, $v = \alpha$, where complex α and real β are infinitesimal parameters. In this case the compensating $U(1)$ transformation is $h(x)$ in (5.17) with the infinitesimal parameter

$$\Sigma = (\beta - \operatorname{Im} \alpha) \frac{1 - |B|^2}{|1 + B|^2} . \tag{5.37}$$

The infinitesimal $SU(1, 1)$ transformation with this compensating $U(1)$ transformation added becomes $\delta B = \alpha + 2i\beta B - \alpha^* B^2$. The corresponding finite $SU(1, 1)$ transformation is given by

$$B \to B' = \frac{uB + v}{v^* B + u^*} \qquad \left(|u|^2 - |v|^2 = 1 \right) . \tag{5.38}$$

To contact with superstring theory it is convenient to make a change of variables from the complex scalar field $B(x)$ satisfying $|B(x)| < 1$ to a complex scalar field $\tau(x)$ taking values in the upper half complex plane $\operatorname{Im} \tau > 0$ by

$$\tau = i \frac{1 - B}{1 + B} . \tag{5.39}$$

From (5.38) the $SU(1, 1)$ transformation of τ can be written as a linear fractional transformation of $SL(2, \mathbb{R})$ ($\sim SU(1, 1)$)

$$\tau \to \tau' = \frac{a\tau + b}{c\tau + d} \qquad (a, b, c, d \in \mathbb{R}, \ ad - bc = 1) . \tag{5.40}$$

Expressing the complex fields $\tau(x)$ and $B^\alpha_{\mu\nu}(x)$ in terms of the real fields $C(x), \phi(x)$, $B^{(i)}_{\mu\nu}(x)$ $(i = 1, 2)$ as

$$\tau = C + \mathrm{i}\,\mathrm{e}^{-\phi}\,,\qquad B^1_{\mu\nu} = \frac{1}{\sqrt{2}}\left(B^{(1)}_{\mu\nu} + \mathrm{i}B^{(2)}_{\mu\nu}\right) = (B^2_{\mu\nu})^* \qquad (5.41)$$

the Lagrangian (5.34) becomes

$$\mathscr{L} = eR - \frac{1}{2}e\left(\partial_\mu\phi\partial^\mu\phi + \mathrm{e}^{2\phi}\partial_\mu C\partial^\mu C\right) - \frac{1}{12}e\,\mathrm{e}^{-\phi}F^{(1)}_{\mu\nu\rho}F^{(1)\mu\nu\rho}$$

$$- \frac{1}{12}e\,\mathrm{e}^{\phi}(F^{(2)} - CF^{(1)})_{\mu\nu\rho}(F^{(2)} - CF^{(1)})^{\mu\nu\rho} - \frac{1}{4\times 5!}e F_{\mu\nu\rho\sigma\tau}F^{\mu\nu\rho\sigma\tau}$$

$$+ \frac{1}{144\times 4!}\varepsilon^{\mu_1\dots\mu_{10}}B_{\mu_1\dots\mu_4}\varepsilon_{ij}F^{(i)}_{\mu_5\mu_6\mu_7}F^{(j)}_{\mu_8\mu_9\mu_{10}} + \cdots\,, \qquad (5.42)$$

where $\cdots$ denote terms depending on the fermionic fields, and $F^{(i)}_{\mu\nu\rho} = 3\partial_{[\mu}B^{(i)}_{\nu\rho]}$, $F_{\mu\nu\rho\sigma\tau} = 5\partial_{[\mu}B_{\nu\rho\sigma\tau]} - 5\varepsilon_{ij}B^{(i)}_{[\mu\nu}F^{(j)}_{\rho\sigma\tau]}$ with $\varepsilon_{ij} = -\varepsilon_{ji}$, $\varepsilon_{12} = +1$.

As for the local supertransformation the parameter of the compensating $U(1)$ transformation is

$$\Sigma = \frac{1}{4\sqrt{2}}\mathrm{i}\left(\bar{\varepsilon}^c\lambda - \bar{\lambda}\varepsilon^c\right)\,. \qquad (5.43)$$

With this compensating transformation the local supertransformation of the scalar field $\tau(x)$ becomes

$$\delta\tau = -\frac{1}{\sqrt{2}}\mathrm{i}\,\mathrm{e}^{-\phi}\bar{\varepsilon}^c\lambda\,. \qquad (5.44)$$

The supertransformations of $\psi_\mu(x)$ and $\lambda(x)$ are also modified by the compensating $U(1)$ transformation (5.43) but the modifications are higher order terms in the fermionic fields, which we omitted in (5.26).

String Frame

$D = 10$, $\mathscr{N} = (2,0)$ supergravity is a low energy effective theory of type IIB superstring theory. The string frame Lagrangian is obtained from (5.42) in the Einstein frame by the Weyl transformation (5.14) as

$$\mathscr{L} = e\,\mathrm{e}^{-2\phi}\left[R + 4\partial_\mu\phi\partial^\mu\phi - \frac{1}{12}F^{(1)}_{\mu\nu\rho}F^{(1)\mu\nu\rho}\right] - \frac{1}{4\times 5!}e F_{\mu\nu\rho\sigma\tau}F^{\mu\nu\rho\sigma\tau}$$

$$- \frac{1}{2}e\partial_\mu C\partial^\mu C - \frac{1}{12}e(F^{(2)} - CF^{(1)})_{\mu\nu\rho}(F^{(2)} - CF^{(1)})^{\mu\nu\rho}$$

$$+ \frac{1}{144\times 4!}\varepsilon^{\mu_1\dots\mu_{10}}B_{\mu_1\dots\mu_4}\varepsilon_{ij}F^{(i)}_{\mu_5\mu_6\mu_7}F^{(j)}_{\mu_8\mu_9\mu_{10}} + \cdots\,, \qquad (5.45)$$

where $\cdots$ denote terms depending on the fermionic fields. The terms depending only on the fields $e_\mu{}^a$, ϕ, $B^{(1)}_{\mu\nu}$ in the NS–NS sector of superstring have the dilaton factor $\mathrm{e}^{-2\phi}$ while the terms depending also on the fields C, $B^{(2)}_{\mu\nu}$, $B_{\mu\nu\rho\sigma}$ in the R–R sector do not. The self-duality equation (5.29) has the same form in both of the Einstein frame and the string frame.

It was conjectured that type IIB superstring theory has a symmetry $SL(2, \mathbb{Z})$, which is a discrete subgroup of $SL(2, \mathbb{R})$ [9]. The scalar field transforms under $SL(2, \mathbb{Z})$ as in (5.40) with integer a, b, c, d. It contains a transformation $\tau \to -1/\tau$. Since the dilaton factor e^ϕ represents the string coupling constant g, this transformation exchanges a strong coupling $g \gg 1$ and a weak coupling $g \ll 1$. Such a relation connecting a strong coupling theory and a weak coupling theory is called the S duality.

5.5 $D = 10, \mathcal{N} = (1, 0)$ Poincaré Supergravity

$D = 10, \mathcal{N} = (1, 0)$ Poincaré supergravity [1, 3, 4] is a low energy effective theory of type I superstring and heterotic string theories. This theory is a chiral theory, which is asymmetric in positive and negative chiralities. The supergravity multiplet consists of a gravitational field $e_\mu{}^a(x)$, a Majorana–Weyl Rarita–Schwinger field of positive chirality $\psi_\mu(x)$, a real antisymmetric tensor field $B_{\mu\nu}(x)$, a Majorana–Weyl spinor field of negative chirality $\lambda(x)$ and a real scalar field $\phi(x)$. Gauge (super Yang–Mills) multiplets can be coupled to this supergravity multiplet as matter supermultiplets. The gauge multiplet consists of a Yang–Mills field $A_\mu(x)$ and a Majorana–Weyl spinor field of positive chirality $\chi(x)$, both of which belong to the adjoint representation of the gauge group.

Pure $D = 10, \mathcal{N} = (1, 0)$ supergravity consisting of only the supergravity multiplet can be obtained from $\mathcal{N} = (1, 1)$ or $\mathcal{N} = (2, 0)$ supergravity by truncations as in Sect. 2.7. To obtain it from $\mathcal{N} = (1, 1)$ supergravity in Sect. 5.3 we impose the conditions $\bar{\gamma}\psi_\mu = \psi_\mu$, $\bar{\gamma}\lambda = -\lambda$, $B_{\mu\nu\rho} = 0$, $B_\mu = 0$ in the $\mathcal{N} = (1, 1)$ theory. These conditions are consistent with the field equations and the local supertransformation with a parameter satisfying $\bar{\gamma}\varepsilon = \varepsilon$. Similarly, to obtain it from $\mathcal{N} = (2, 0)$ supergravity in Sect. 5.4 we impose the conditions $\psi_\mu^c = \psi_\mu$, $\lambda^c = \lambda$, $B_{\mu\nu\rho\sigma} = 0$, $B_{\mu\nu}^{(2)} = 0, C = 0$ and make the replacements $\psi_\mu \to \frac{1}{\sqrt{2}}\psi_\mu, \lambda \to \frac{1}{\sqrt{2}}\lambda, B_{\mu\nu}^{(1)} \to B_{\mu\nu}$. These conditions are consistent with the field equations and the local supertransformation with a parameter satisfying $\varepsilon^c = \varepsilon$. These two truncations give the same $\mathcal{N} = (1, 0)$ theory.

The Lagrangian of the super Yang–Mills theory in a flat Minkowski spacetime is

$$\mathscr{L} = \frac{1}{2g^2}\text{tr}\left(F_{\mu\nu}F^{\mu\nu}\right) + \frac{1}{g^2}\text{tr}\left(\bar{\chi}\gamma^\mu D_\mu\chi\right) . \tag{5.46}$$

Here, we have used matrix-valued fields $A_\mu(x) = -igA_\mu^I(x)T_I$, $\chi(x) = -ig\chi^I(x)$ T_I, where g is a gauge coupling constant and T_I are representation matrices of the Lie algebra of the gauge group (See Sect. 1.3). The field strength and the covariant derivatives are given by

$$F_{\mu\nu} = \partial_\mu A_\nu - \partial_\nu A_\mu + [A_\mu, A_\nu] , \quad D_\mu\chi = \partial_\mu\chi + [A_\mu, \chi] . \tag{5.47}$$

Table 5.3 $\mathbb{R}^+$ weights in $D = 10$, $\mathcal{N} = (1, 0)$ supergravity

Field	$e_\mu{}^a$	$B_{\mu\nu}$	e^ϕ	ψ_μ	λ	A_μ^I	χ^I
w	0	$\frac{1}{2}$	1	0	0	$\frac{1}{4}$	0

This Lagrangian is invariant up to total divergences under the global $\mathcal{N} = (1, 0)$ supertransformation

$$\delta_Q A_\mu = -\frac{1}{2\sqrt{2}}\bar{\varepsilon}\gamma_\mu\chi \, , \qquad \delta_Q\chi = \frac{1}{4\sqrt{2}}\gamma^{\mu\nu}\varepsilon F_{\mu\nu} \, , \tag{5.48}$$

where the transformation parameter ε is a constant Majorana–Weyl spinor of positive chirality.

The Lagrangian of the super Yang–Mills theory coupled to the supergravity is

$$
\begin{aligned}
\mathscr{L} &= eR - \frac{1}{2}e\partial_\mu\phi\partial^\mu\phi - \frac{1}{12}e\,\mathrm{e}^{-\phi}F_{\mu\nu\rho}F^{\mu\nu\rho} - \frac{1}{2}e\bar{\psi}_\mu\gamma^{\mu\nu\rho}D_\nu\psi_\rho - \frac{1}{2}e\bar{\lambda}\gamma^\mu D_\mu\lambda \\
&\quad + \frac{1}{2g^2}e\,\mathrm{e}^{-\frac{1}{2}\phi}\mathrm{tr}\left(F_{\mu\nu}F^{\mu\nu}\right) + \frac{1}{g^2}e\,\mathrm{tr}\left(\bar{\chi}\gamma^\mu D_\mu\chi\right) - \frac{1}{2\sqrt{2}}e\bar{\psi}_\mu\gamma^\nu\gamma^\mu\lambda\,\partial_\nu\phi \\
&\quad + \frac{1}{48}e\,\mathrm{e}^{-\frac{1}{2}\phi}\left[\bar{\psi}_{[\mu}\gamma^\mu\gamma^{\rho\sigma\tau}\gamma^\nu\psi_{\nu]} - \sqrt{2}\bar{\psi}_\mu\gamma^{\rho\sigma\tau}\gamma^\mu\lambda - \frac{2}{g^2}\mathrm{tr}\left(\bar{\chi}\gamma^{\rho\sigma\tau}\chi\right)\right]F_{\rho\sigma\tau} \\
&\quad - \frac{1}{2\sqrt{2}g^2}e\,\mathrm{e}^{-\frac{1}{4}\phi}\mathrm{tr}\left[\left(\bar{\psi}_\mu\gamma^{\rho\sigma}\gamma^\mu\chi - \frac{1}{\sqrt{2}}\bar{\lambda}\gamma^{\rho\sigma}\chi\right)F_{\rho\sigma}\right] + \cdots \, ,
\end{aligned}
\tag{5.49}
$$

where $\cdots$ denote terms higher order in the fermionic fields. The covariant derivative on χ contains the Yang–Mills field as in (5.47) in addition to the spin connection. The field strength of the Yang–Mills field and the antisymmetric tensor field are given by (5.47) and

$$F_{\mu\nu\rho} = 3\partial_{[\mu}B_{\nu\rho]} + \frac{6}{g^2}\mathrm{tr}\left(A_{[\mu}\partial_\nu A_{\rho]} + \frac{2}{3}A_{[\mu}A_\nu A_{\rho]}\right) \, , \tag{5.50}$$

which contains a term depending on the Yang–Mills field. This term is called the Chern–Simons term and plays an important role in the Green–Schwarz mechanism of anomaly cancellations [7].

The Lagrangian (5.49) is invariant under the global $\mathbb{R}^+$ transformation $\Phi(x) \to \mathrm{e}^{w\Lambda}\Phi(x)$, where the weight w of each field $\Phi(x)$ is given in Table 5.3. It is also invariant up to total divergences under the general coordinate transformation $\delta_G(\xi)$, the local Lorentz transformation $\delta_L(\lambda)$, the gauge transformation of the antisymmetric tensor field $\delta_g(\zeta)$, the Yang–Mills gauge transformation $\delta_{\mathrm{YM}}(v)$ and the local supertransformation $\delta_Q(\varepsilon)$. The gauge transformations of the antisymmetric tensor field and the Yang–Mills field are

$$(\delta_g + \delta_{\mathrm{YM}}) B_{\mu\nu} = 2\partial_{[\mu}\zeta_{\nu]} - \frac{2}{g^2}\mathrm{tr}\left(v\partial_{[\mu}A_{\nu]}\right) , \qquad (5.51)$$

and (1.34). The field strength (5.50) is invariant under these gauge transformations.

The local supertransformation with a Majorana–Weyl spinor parameter $\varepsilon(x)$ of positive chirality is

$$\delta_Q e_\mu{}^a = \frac{1}{4}\bar{\varepsilon}\gamma^a\psi_\mu , \qquad \delta_Q\phi = -\frac{1}{2\sqrt{2}}\bar{\varepsilon}\lambda ,$$

$$\delta_Q B_{\mu\nu} = \frac{1}{2}e^{\frac{1}{2}\phi}\bar{\varepsilon}\gamma_{[\mu}\psi_{\nu]} - \frac{1}{4\sqrt{2}}e^{\frac{1}{2}\phi}\bar{\varepsilon}\gamma_{\mu\nu}\lambda + \frac{1}{\sqrt{2}g^2}e^{\frac{1}{4}\phi}\mathrm{tr}\left(\bar{\varepsilon}\gamma_{[\mu}\chi A_{\nu]}\right) ,$$

$$\delta_Q\psi_\mu = D_\mu\varepsilon + \frac{1}{96}e^{-\frac{1}{2}\phi}\left(\gamma_\mu{}^{\nu\rho\sigma} - 9\delta_\mu^\nu\gamma^{\rho\sigma}\right)\tilde{\gamma}\varepsilon F_{\nu\rho\sigma} + \cdots ,$$

$$\delta_Q\lambda = -\frac{1}{2\sqrt{2}}\gamma^\mu\varepsilon\partial_\mu\phi + \frac{1}{24\sqrt{2}}e^{-\frac{1}{2}\phi}\gamma^{\mu\nu\rho}\varepsilon F_{\mu\nu\rho} + \cdots ,$$

$$\delta_Q A_\mu = -\frac{1}{2\sqrt{2}}e^{\frac{1}{4}\phi}\bar{\varepsilon}\gamma_\mu\chi , \qquad \delta_Q\chi = \frac{1}{4\sqrt{2}}e^{-\frac{1}{4}\phi}\gamma^{\mu\nu}\varepsilon F_{\mu\nu} + \cdots , \qquad (5.52)$$

where $\cdots$ denote terms higher order in the fermionic fields. The commutator of two supertransformations is

$$[\delta_Q(\varepsilon_1), \delta_Q(\varepsilon_2)] = \delta_G(\xi) + \delta_L(\lambda) + \delta_g(\zeta) + \delta_{\mathrm{YM}}(v) + \delta_Q(\varepsilon) . \qquad (5.53)$$

The transformation parameters on the right-hand side are

$$\xi^\mu = \frac{1}{4}\bar{\varepsilon}_2\gamma^\mu\varepsilon_1 , \quad \zeta_\mu = -\xi^\nu B_{\nu\mu} - e^{\frac{1}{2}\phi}\xi_\mu + \cdots , \quad v = -\xi^\mu A_\mu + \cdots , \quad \varepsilon = \cdots ,$$

$$\lambda_{ab} = -\xi^\mu\omega_{\mu ab} - \frac{1}{192}e^{-\frac{1}{2}\phi}\left(\bar{\varepsilon}_2\gamma_{ab}{}^{\mu\nu\rho}\varepsilon_1 F_{\mu\nu\rho} + 72\xi^\mu F_{\mu ab}\right) + \cdots , \qquad (5.54)$$

where $\cdots$ denote terms depending on the fermionic fields. To derive this commutation relation we have to use the field equations of the fermionic fields.

String Frame

$D = 10$, $\mathscr{N} = (1, 0)$ supergravity is a low energy effective theory of type I superstring and heterotic string theories. The effective Lagrangians of these string theories are different from (5.49). The effective Lagrangian of type I superstring is

$$\mathscr{L} = e\,e^{-2\phi}\left(R + 4\partial_\mu\phi\partial^\mu\phi\right) - \frac{1}{12}eF_{\mu\nu\rho}F^{\mu\nu\rho} + \frac{1}{2g^2}e\,e^{-\phi}\mathrm{tr}\left(F_{\mu\nu}F^{\mu\nu}\right) + \cdots , \qquad (5.55)$$

where $\cdots$ denote terms depending on the fermionic fields. The first term depending only on the fields $e_\mu{}^a$, ϕ in the NS–NS sector contains the dilaton factor $e^{-2\phi}$, but the second term depending also on the field $B_{\mu\nu}$ in the R–R sector does not. The third term depending on the Yang–Mills field A_μ, which is an open string state, has the factor $e^{-\phi}$. On the other hand, the effective Lagrangian of heterotic string is

$$\mathscr{L} = e\, e^{-2\phi}\left[R + 4\partial_\mu\phi\partial^\mu\phi - \frac{1}{12}F_{\mu\nu\rho}F^{\mu\nu\rho} + \frac{1}{2g^2}\operatorname{tr}\left(F_{\mu\nu}F^{\mu\nu}\right)\right] + \cdots . \quad (5.56)$$

In this case all the terms have the dilaton factor $e^{-2\phi}$. These two Lagrangians in the string frame can be obtained from (5.49) in the Einstein frame by a change of the field variables

$$\text{Type I: } e_\mu{}^a \to e^{-\frac{1}{4}\phi}e_\mu{}^a, \quad \phi \to -\phi ,$$

$$\text{heterotic: } e_\mu{}^a \to e^{-\frac{1}{4}\phi}e_\mu{}^a, \quad \phi \to \phi . \quad (5.57)$$

When the gauge group is $SO(32)$, there exist both of type I superstring theory and heterotic string theory. Since both of their effective theories can be derived from the Lagrangian (5.49), there is a relation between them. In fact, if we make a change of the variables $e_\mu{}^a \to e^{-\frac{1}{2}\phi}e_\mu{}^a, \phi \to -\phi$ in (5.55), then we obtain (5.56). It was conjectured that type I superstring theory and heterotic string theory with the gauge group $SO(32)$ are equivalent by the S duality [11]. The fact that the effective theories of these two string theories are related by a change of variables is an evidence for this S duality. By this change of the variables the string coupling constant e^ϕ transforms to its inverse $e^{-\phi}$. Therefore, strongly (weakly) coupled type I superstring theory corresponds to weakly (strongly) coupled heterotic string theory.

References

1. E. Bergshoeff, M. de Roo, B. de Wit, P. van Nieuwenhuizen, Ten-dimensional Maxwell–Einstein supergravity, its currents, and the issue of its auxiliary fields. Nucl. Phys. **B195**, 97 (1982)
2. I.C.G. Campbell, P.C. West, $\mathcal{N} = 2$, $D = 10$ nonchiral supergravity and its spontaneous compactification. Nucl. Phys. **B243**, 112 (1984)
3. A.H. Chamseddine, $\mathcal{N} = 4$ supergravity coupled to $\mathcal{N} = 4$ matter. Nucl. Phys. **B185**, 403 (1981)
4. G.F. Chapline, N.S. Manton, Unification of Yang-Mills theory and supergravity in ten-dimensions. Phys. Lett. **B120**, 105 (1983)
5. E. Cremmer, B. Julia, J. Scherk, Supergravity theory in 11 dimensions. Phys. Lett. **B76**, 409 (1978)
6. F. Giani, M. Pernici, $\mathcal{N} = 2$ supergravity in ten-dimensions. Phys. Rev. **D30**, 325 (1984)
7. M.B. Green, J.H. Schwarz, Anomaly cancellation in supersymmetric $D = 10$ gauge theory and superstring theory. Phys. Lett. **B149**, 117 (1984)
8. P.S. Howe, P.C. West, The complete $\mathcal{N} = 2$, $D = 10$ supergravity. Nucl. Phys. **B238**, 181 (1984)
9. C.M. Hull, P.K. Townsend, Unity of superstring dualities. Nucl. Phys. **B438**, 109 (1995). [hep-th/9101]
10. M. Huq, M.A. Namazie, Kaluza–Klein supergravity in ten-dimensions. Class. Quant. Grav. **2**, 293 (1985). [Erratum-ibid. 2 (1985) 597]

11. J. Polchinski, E. Witten, Evidence for heterotic–type I string duality. Nucl. Phys. **B460**, 525 (1996). [hep-th/9510169]
12. A. Salam, E. Sezgin, *Supergravities in Diverse Dimensions* (World Scientific, North-Holland, 1989)
13. J.H. Schwarz, Covariant field equations of chiral $\mathcal{N} = 2$ $D = 10$ supergravity. Nucl. Phys. **B226**, 269 (1983)

Chapter 6
Dimensional Reductions

6.1 Compactifications and Dimensional Reductions

Since superstring and M theories are formulated in spacetime of 10 or 11 dimensions, one has to find a method to obtain four-dimensional theories from higher dimensions. The compactification is one of such methods, in which one assumes that some of the dimensions of spacetime form a small compact space.

As a simple example of compactifications let us consider a theory of a real scalar field ϕ in five dimensions. We assume that five-dimensional spacetime is a direct product $M_4 \times S^1$ of four-dimensional Minkowski spacetime M_4 and a circle S^1 of radius R. The action of the five-dimensional theory with a ϕ^4 interaction is

$$S = \int d^4x dy \left[-\frac{1}{2}\eta^{MN}\partial_M\phi\partial_N\phi - \frac{1}{4!}\lambda_5\phi^4 \right], \tag{6.1}$$

where we have split five-dimensional coordinates x^M ($M = 0, 1, 2, 3, 4$) into four-dimensional coordinates x^μ ($\mu = 0, 1, 2, 3$) of M_4 and a coordinate $x^4 = y$ ($0 \leq y \leq 2\pi R$) of S^1. Since the field is periodic in the y direction with period $2\pi R$, it can be expanded in a Fourier series as

$$\phi(x, y) = \frac{1}{\sqrt{2\pi R}} \sum_{n=-\infty}^{\infty} \phi_n(x)\, e^{iny/R}, \qquad (\phi_n)^* = \phi_{-n}. \tag{6.2}$$

The expansion coefficients $\phi_n(x)$ ($n \in \mathbb{Z}$) are functions of the four-dimensional coordinates x^μ. Substituting this into the action and integrating over y we obtain

Y. Tanii, *Introduction to Supergravity*, SpringerBriefs in Mathematical Physics, DOI: 10.1007/978-4-431-54828-7_6, © The Author(s) 2014

$$S = \int d^4x \left[-\frac{1}{2}\partial_\mu\phi_0\partial^\mu\phi_0 - \frac{1}{4!}\lambda_4\phi_0^4 - \sum_{n=1}^{\infty}\left\{\partial_\mu\phi_n^*\partial^\mu\phi_n + \left(\frac{n}{R}\right)^2\phi_n^*\phi_n\right\}\right.$$
$$\left. -\frac{1}{4!}\lambda_4 \sum_{n_1,n_2,n_3=-\infty}^{\infty}{}' \phi_{n_1}\phi_{n_2}\phi_{n_3}\phi_{-n_1-n_2-n_3}\right], \tag{6.3}$$

where $\lambda_4 = \lambda_5/(2\pi R)$ and the sum $\sum'$ excludes the term $(n_1, n_2, n_3) = (0, 0, 0)$. From this action we see that the zero mode $\phi_0(x)$ represents a massless real scalar field and the higher modes $\phi_n(x)$ ($n = 1, 2, 3, \ldots$) represent massive complex scalar fields of mass $m_n = n/R$ in four dimensions. These massive higher modes are called Kaluza–Klein modes. In this way the five-dimensional theory reduces· to a four-dimensional theory which contains infinitely many fields. Since the Kaluza–Klein modes have masses of order $1/R$, they can be ignored at sufficiently low energy compared to $1/R$. Since the terms in the action (6.3) depending on the Kaluza–Klein modes are at least quadratic in them, we can set $\phi_n = 0$ ($n \neq 0$) consistently with their field equations. As a consequence, we obtain a four-dimensional field theory which contains only the massless zero mode $\phi_0(x)$ at low energy.

In this example the compact space is a circle S^1. In general, one can consider compactifications using other compact spaces of various dimensions. For instance, one can use a d-dimensional torus T^d or a d-dimensional sphere S^d. In string theory and M theory, compactifications using special compact spaces such as Calabi–Yau manifolds and G_2 holonomy manifolds are interesting from a phenomenological point of view. In this book we will consider only troidal compactifications, i.e., compactifications using tori. For more general compactifications, see the textbooks on string theory referred to at the end of Sect. 1.1.

In a theory containing gravity, spacetime is determined by the gravitational field equation. Therefore, a compact space should be chosen so that the total spacetime is a solution of the field equations. Such a compactification in which a compact space appears as a consequence of the field equations is called the spontaneous compactification.

Dimensional Reductions

One can obtain a lower dimensional theory of the zero modes by discarding massive Kaluza–Klein modes in a toroidal compactification of a higher dimensional theory. Such a mathematical procedure is called the dimensional reduction. Discarding massive modes is equivalent to assuming that higher dimensional fields do not depend on the coordinates of the torus. For instance, to dimensionally reduce the five-dimensional theory (6.1) to four dimensions, we only need to assume that the field is independent of y and rescale it as $\phi(x, y) \rightarrow \phi_0(x)/\sqrt{2\pi R}$.

Dimensional reductions of multi-component fields such as vector fields and a gravitational field show more complex structure. Let us consider a dimensional reduction of a five-dimensional theory of a vector field to four dimensions. The five-dimensional vector field has components $(A_\mu(x), A_y(x))$ ($\mu = 0, 1, 2, 3$). The components A_μ

in the uncompactified directions represent a vector field while the component A_y in the compactified direction represents a scalar field in four dimensions. Similarly, a five dimensional gravitational field gives a gravitational field $g_{\mu\nu}(x)$, a vector field $g_{\mu y}(x)$ and a scalar field $g_{yy}(x)$ in four dimensions. To obtain the standard form of the Lagrangian in the lower dimensional theory we have to make appropriate field redefinitions in lower dimensions as we will discuss below.

By a dimensional reduction of a supergravity in higher dimensions we can obtain a supergravity in lower dimensions. Since the number of supersymmetries is preserved by a troidal compactification and a truncation to the zero mode sector, the lower dimensional supergravity has the same number of supersymmetries as the higher dimensional one. (For more general compactifications such as those using Calabi–Yau manifolds, the number of supersymmetries decreases in general.) Supergravities generically contain scalar fields and their Lagrangians are non-polynomials in the scalar fields. However, $D = 11$ supergravity does not contain scalar fields and has a relatively simple form. Hence, it is convenient to start from the $D = 11$ supergravity and dimensionally reduce it in order to obtain supergravities in lower dimensions. In fact, $D = 4$, $\mathcal{N} = 8$ supergravity discussed in Sect. 4.3 was first constructed by a dimensional reduction of $D = 11$, $\mathcal{N} = 1$ supergravity [3]. Dimensional reductions are also useful to understand the origins of the global symmetries of supergravities. In the following we consider dimensional reductions of supergravities in 11 and 10 dimensions and see how global symmetries in lower dimensions appear.

6.2 Dimensional Reductions of Field Theories

We first discuss dimensional reductions of various kinds of fields in higher dimensions [14]. We consider dimensional reductions from $\hat{D} = D + d$ dimensions to D dimensions. We denote $\hat{D}$-dimensional coordinates by x^M ($M = 0, 1, \ldots, \hat{D} - 1$), D-dimensional coordinates by x^μ ($\mu = 0, 1, \ldots, D - 1$) and d-dimensional coordinates of the torus by y^α ($\alpha = 1, 2, \ldots, d$). Local Lorentz indices are denoted by $A, B, \ldots = 0, 1, \ldots, \hat{D} - 1$; $a, b, \ldots = 0, 1, \ldots, D - 1$ and $i = 1, 2, \ldots, d$, respectively. The coordinates of the torus are assumed to take values in $0 \leq y^\alpha \leq 1$. In the previous section y takes values in $0 \leq y \leq 2\pi R$, which depends on the radius R of S^1. In the following the size and the shape of the torus are specified by the metric rather than the range of the coordinates. Therefore, D-dimensional action is given by

$$S = \int d^D x \, d^d y \, \mathcal{L} = \int d^D x \, \mathcal{L}, \tag{6.4}$$

where $\mathcal{L}$ is a $\hat{D}$-dimensional Lagrangian for the fields independent of y^α.

6.2.1 Gravitational Field

Let us first consider the dimensional reduction of a gravitational field. We denote the $\hat{D}$-dimensional vielbein as $\hat{e}_M{}^A$. We use $\hat{}$ to represent quantities in $\hat{D}$ dimensions. The Lagrangian in $\hat{D}$ dimensions is

$$\mathscr{L} = \hat{e}\hat{R}. \tag{6.5}$$

We assume that the vielbein $\hat{e}_M{}^A$ does not depend on the coordinates y^α of the torus. Components of $\hat{e}_M{}^A$ represent various kinds of D-dimensional fields. $\hat{e}_\mu{}^a$ represents a gravitational field, $\hat{e}_\mu{}^i$ vector fields and $\hat{e}_\alpha{}^i$ scalar fields. To obtain the standard form of the Lagrangian in D dimensions we have to make some field redefinitions. We express the $\hat{D}$-dimensional vielbein $\hat{e}_M{}^A$ and its inverse matrix $\hat{e}_A{}^M$ in terms of D-dimensional fields as

$$\hat{e}_M{}^A = \begin{pmatrix} e^\sigma e_\mu{}^a & A_\mu{}^\beta E_\beta{}^i \\ 0 & E_\alpha{}^i \end{pmatrix}, \qquad \hat{e}_A{}^M = \begin{pmatrix} e^{-\sigma} e_a{}^\mu & -e^{-\sigma} e_a{}^\nu A_\nu{}^\alpha \\ 0 & E_i{}^\alpha \end{pmatrix}. \tag{6.6}$$

By using a part of $\hat{D}$-dimensional local Lorentz transformations we have set $\hat{e}_\alpha{}^a = 0$. $e_\mu{}^a, A_\mu{}^\alpha, E_\alpha{}^i$ depend only on the D-dimensional coordinates and represent a vielbein, vector fields and scalar fields in D dimensions. $e_a{}^\mu$ and $E_i{}^\alpha$ are the inverse matrices of $e_\mu{}^a$ and $E_\alpha{}^i$, respectively. The factor e^σ defined by

$$e^\sigma = E^{-\frac{1}{D-2}}, \qquad E = \det E_\alpha{}^i \tag{6.7}$$

has been introduced so that the Lagrangian of the gravitational field becomes the Einstein type. The $\hat{D}$-dimensional metric $\hat{g}_{MN} = \hat{e}_M{}^A \hat{e}_N{}^B \eta_{AB}$ is given by

$$dx^M dx^N \hat{g}_{MN} = dx^\mu dx^\nu g_{\mu\nu} e^{2\sigma} + (dy^\alpha + dx^\mu A_\mu{}^\alpha)(dy^\beta + dx^\nu A_\nu{}^\beta) G_{\alpha\beta}, \tag{6.8}$$

where $g_{\mu\nu} = e_\mu{}^a e_\nu{}^b \eta_{ab}$, $G_{\alpha\beta} = E_\alpha{}^i E_\beta{}^j \delta_{ij}$.

6.2.1.1 Lagrangian

Substituting (6.6) into (6.5) we find the D-dimensional Lagrangian as

$$\mathscr{L} = eR - \frac{1}{D-2} eE^{-2} \partial_\mu E \partial^\mu E + \frac{1}{4} e \partial_\mu G^{\alpha\beta} \partial^\mu G_{\alpha\beta} - \frac{1}{4} eE^{\frac{2}{D-2}} F_{\mu\nu}{}^\alpha F^{\mu\nu\beta} G_{\alpha\beta}. \tag{6.9}$$

To rewrite this Lagrangian we introduce $v_\alpha{}^i$ and $S_{\alpha\beta}$ defined by

$$v_\alpha{}^i = E^{-\frac{1}{d}} E_\alpha{}^i, \qquad S_{\alpha\beta} = v_\alpha{}^i v_\beta{}^j \delta_{ij}, \tag{6.10}$$

which satisfy $\det v_\alpha{}^i = 1$, $\det S_{\alpha\beta} = 1$. If we further define the scalar field $\phi(x)$ and the constant a by

$$E = e^{\frac{1}{2}(D-2)da\phi}, \qquad a = \sqrt{\frac{2}{(D-2)d(D+d-2)}}, \tag{6.11}$$

we obtain

$$\mathscr{L} = eR - \frac{1}{2}e\partial_\mu\phi\partial^\mu\phi + \frac{1}{4}e\partial_\mu S^{\alpha\beta}\partial^\mu S_{\alpha\beta} - \frac{1}{4}e\,e^{(D+d-2)a\phi}F_{\mu\nu}{}^\alpha F^{\mu\nu\beta}S_{\alpha\beta}. \tag{6.12}$$

This is a Lagrangian of a gravitational field $g_{\mu\nu}$, $\frac{1}{2}d(d+1)$ scalar fields ϕ, $S_{\alpha\beta}$ and d vector fields $A_\mu{}^\alpha$ in D dimensions. The scalar fields have a non-polynomial kinetic term. As we will see later the scalar fields $S_{\alpha\beta}$ are described by the $SL(d,\mathbb{R})/SO(d)$ non-linear sigma model. From $G_{\alpha\beta} = e^{(D-2)a\phi}S_{\alpha\beta}$ we see that the factor $e^{\frac{1}{2}(D-2)a\phi}$ represents the size of the torus.

6.2.1.2 Symmetries

The Lagrangian in $\hat{D}$ dimensions is invariant up to total divergences under $\hat{D}$-dimensional general coordinate and local Lorentz transformations

$$\delta\hat{e}_M{}^A = \xi^N\partial_N\hat{e}_M{}^A + \partial_M\xi^N\hat{e}_N{}^A - \lambda^A{}_B\hat{e}_M{}^B. \tag{6.13}$$

Let us see how these transformations look like in D dimensions. To be consistent with the dimensional reduction the above transformation must satisfy certain conditions. First, (i) the fields after the transformations must be independent of the coordinates of the torus y^α. Secondly, the above transformations are not necessarily a symmetry of the D-dimensional action since the D-dimensional action is an integral over only D dimensions. Under the transformations (6.13) the Lagrangian transforms as

$$\delta\mathscr{L} = \partial_M\left(\xi^M\mathscr{L}\right) = \partial_\mu\left(\xi^\mu\mathscr{L}\right) + \partial_\alpha\xi^\alpha\mathscr{L}, \tag{6.14}$$

where we have used $\partial_\alpha\mathscr{L} = 0$. When integrating over the D-dimensional coordinates the first term vanishes while the second term does not. Therefore, we need to require (ii) $\partial_\alpha\xi^\alpha = 0$. Furthermore, $\lambda_{ai} = 0$ is needed to preserve $\hat{e}_\alpha{}^a = 0$ in (6.6).

When the transformation parameters ξ^μ, ξ^α, λ_{ab}, λ_{ij} do not depend on the coordinates y^α, the above conditions (i), (ii) are automatically satisfied and we find D-dimensional transformations

$$\delta e_\mu{}^a = \xi^\nu \partial_\nu e_\mu{}^a + \partial_\mu \xi^\nu e_\nu{}^a - \lambda^a{}_b e_\mu{}^b,$$

$$\delta A_\mu{}^\alpha = \xi^\nu \partial_\nu A_\mu{}^\alpha + \partial_\mu \xi^\nu A_\nu{}^\alpha + \partial_\mu \xi^\alpha,$$

$$\delta\phi = \xi^\nu \partial_\nu \phi, \qquad \delta v_\alpha{}^i = \xi^\nu \partial_\nu v_\alpha{}^i + v_\alpha{}^j \lambda_j{}^i. \qquad (6.15)$$

ξ^μ and λ_{ab} represent the general coordinate and local Lorentz transformations. The coordinate transformation in the compactified direction ξ^α (ξ^α transformation) represents a $U(1)^d$ gauge transformation of the vector fields $A_\mu{}^\alpha$. λ_{ij} represents a local $SO(d)$ transformation acting on the scalar fields $v_\alpha{}^i$. D-dimensional action made from the Lagrangian (6.12) is invariant under these transformations.

Transformations linear in the compactified coordinate $\xi^\alpha = y^\beta L_\beta{}^\alpha$, where $L_\beta{}^\alpha$ are constants satisfying $L_\alpha{}^\alpha = 0$, also satisfy the above two conditions (i), (ii). In this case D-dimensional transformations are

$$\delta v_\alpha{}^i = L_\alpha{}^\beta v_\beta{}^i, \quad \delta A_\mu{}^\alpha = -A_\mu{}^\beta L_\beta{}^\alpha, \quad \delta(\text{other fields}) = 0. \qquad (6.16)$$

These transformations represent a global $SL(d, \mathbb{R})$ symmetry.

The transformation $\xi^\alpha = c y^\alpha$ (c = constant) satisfies the condition (i) but not the condition (ii) $\partial_\alpha \xi^\alpha = 0$. Hence, the D-dimensional action is not invariant under it. In fact, we find $\delta\mathcal{L} = cd\mathcal{L}$ from (6.14). However, in the special case of the $\hat{D}$-dimensional Lagrangian (6.5), there is another transformation $\delta \hat{e}_M{}^A = c' \hat{e}_M{}^A$ (c' = constant), under which the Lagrangian transforms as $\delta\mathcal{L} = (D + d - 2)c'\mathcal{L}$. Therefore, under a combination of these two transformations with $c' = -dc/(D + d - 2)$ the Lagrangian is invariant. This combined transformation is

$$\delta\phi = \phi + \frac{2c}{(D+d-2)a}, \quad \delta A_\mu{}^\alpha = -c A_\mu{}^\alpha, \quad \delta(\text{other fields}) = 0, \qquad (6.17)$$

which represents a global $\mathbb{R}^+$ symmetry. The existence of this $\mathbb{R}^+$ symmetry depends on a particular form of the Lagrangian (6.5). Other Lagrangians may not have $\mathbb{R}^+$ symmetry even if it is invariant under the $\hat{D}$-dimensional general coordinate transformation. For instance, when a $\hat{D}$-dimensional Lagrangian has a cosmological term, the D-dimensional action derived from it is not $\mathbb{R}^+$ invariant.

6.2.1.3 $SL(d, \mathbb{R})/SO(d)$ Non-Linear Sigma Model

The scalar fields $v_\alpha{}^i$ are components of a $d \times d$ matrix with unit determinant and can be regarded as representing an element of the group $SL(d, \mathbb{R})$. Its $SO(d)$ subgroup part can be removed by the local $SO(d)$ transformations with parameter λ_{ij}. Therefore, the scalar fields can be described by the $SL(d, \mathbb{R})/SO(d)$ non-linear sigma model. Following Sect. 4.1 we define $P_{\mu ij}$ and $Q_{\mu ij}$ by $v^{-1}\partial_\mu v = P_\mu + Q_\mu$ ($P_\mu^T = P_\mu, Q_\mu^T = -Q_\mu$). Q_μ belongs to the $SO(d)$ Lie subalgebra and P_μ is in its orthogonal complement. Explicitly we find

$$P_{\mu ij} = \frac{1}{2} \left(v_i{}^\alpha \partial_\mu v_{\alpha j} + v_j{}^\alpha \partial_\mu v_{\alpha i} \right), \quad Q_{\mu ij} = \frac{1}{2} \left(v_i{}^\alpha \partial_\mu v_{\alpha j} - v_j{}^\alpha \partial_\mu v_{\alpha i} \right), \quad (6.18)$$

where $v_i{}^\alpha$ is the inverse matrix of $v_\alpha{}^i$. As in (4.5) the kinetic term of $S_{\alpha\beta}$ in the Lagrangian (6.12) can be written as $\frac{1}{4} e \partial_\mu S^{\alpha\beta} \partial^\mu S_{\alpha\beta} = -\frac{1}{2} e \operatorname{tr} \left(P_\mu P^\mu \right)$. The scalar fields $S_{\alpha\beta} = v_\alpha{}^i v_\beta{}^j \delta_{ij}$ are $SO(d)$ invariant variables made from $v_\alpha{}^i$ and represent $\frac{1}{2} d(d+1) - 1$ physical degrees of freedom. Similarly, the scalar field ϕ is described by the $\mathbb{R}^+$ non-linear sigma model.

6.2.2 Yang–Mills Field

The Lagrangian of a Yang–Mills field coupled to the gravitational field in $\hat{D}$ dimensions is

$$\mathscr{L} = \frac{1}{2g^2} \hat{e} \operatorname{tr} \left(\hat{F}_{MN} \hat{F}^{MN} \right), \quad (6.19)$$

where g is a gauge coupling constant. As in Sect. 1.3 we use a matrix-valued gauge field $\hat{A}_M = -ig\hat{A}_M^I T_I$ and its field strength $\hat{F}_{MN} = \partial_M \hat{A}_N - \partial_N \hat{A}_M + [\hat{A}_M, \hat{A}_N]$. This Lagrangian is invariant under the gauge transformation $\delta \hat{A}_M = \hat{D}_M v = \partial_M v + [\hat{A}_M, v]$.

To obtain D-dimensional Lagrangian we first assume that the fields are independent of the coordinates of the torus y^α. The components $\hat{A}_\mu$ and $\hat{A}_\alpha$ become a vector field and d scalar fields, respectively. As we saw in Sect. 6.2.1, the coordinate transformation in the compactified directions becomes the gauge transformation (ξ^α transformation) of the vector fields $A_\mu{}^\alpha$ resulting from the $\hat{D}$-dimensional gravitational field. The gauge field $\hat{A}_\mu$ transforms non-trivially under the ξ^α transformation. It is convenient to define D-dimensional fields so that they are invariant under the ξ^α transformation. We can use the following trick to find such definitions. The $\hat{D}$-dimensional fields with local Lorentz indices $\hat{A}_A = \hat{e}_A{}^M \hat{A}_M$ transform as scalars under the $\hat{D}$-dimensional general coordinate transformation. Since they are independent of the d-dimensional coordinates, they are invariant under the ξ^α transformation. We define the D-dimensional fields A_μ, A_α from $\hat{A}_A$ as

$$A_\mu \equiv e^\sigma e_\mu{}^a \hat{A}_a = \hat{A}_\mu - A_\mu{}^\alpha \hat{A}_\alpha, \quad A_\alpha \equiv E_\alpha{}^i \hat{A}_i = \hat{A}_\alpha. \quad (6.20)$$

Since $e_\mu{}^a$ and $E_\alpha{}^i = e^{\frac{1}{2}(D-2)a\phi} v_\alpha{}^i$ are invariant under the ξ^α transformation as can be seen in (6.15), A_μ and A_α are also invariant. Under the D-dimensional general coordinate transformation in the x^μ directions these fields transform as a vector field and scalar fields, respectively. From the $\hat{D}$-dimensional gauge transformation $\delta \hat{A}_M = \hat{D}_M v$ with a parameter v independent of y^α we find the D-dimensional transformation

$$\delta A_\mu = D_\mu v, \qquad \delta A_\alpha = [A_\alpha, v], \tag{6.21}$$

where D_μ is the covariant derivative using A_μ.

It is also convenient to define the field strengths invariant under the ξ^α transformation from $\hat{F}_{AB} = \hat{e}_A{}^M \hat{e}_B{}^N \hat{F}_{MN}$ as

$$\tilde{F}_{\mu\nu} \equiv e^{2\sigma} e_\mu{}^a e_\nu{}^b \hat{F}_{ab} = F_{\mu\nu} + F_{\mu\nu}{}^\alpha A_\alpha,$$
$$e^\sigma e_\mu{}^a E_\alpha{}^i \hat{F}_{ai} = D_\mu A_\alpha, \qquad E_\alpha{}^i E_\beta{}^j \hat{F}_{ij} = [A_\alpha, A_\beta], \tag{6.22}$$

where $F_{\mu\nu}$ is the field strength of A_μ. These quantities transform covariantly under the gauge transformations (6.21). The Lagrangian (6.19) then becomes

$$\mathscr{L} = \frac{1}{2g^2} e\, e^{da\phi} \mathrm{tr}\left(\tilde{F}_{\mu\nu}\tilde{F}^{\mu\nu}\right) + \frac{1}{g^2} e\, e^{-(D-2)a\phi} \mathrm{tr}\left(D_\mu A_\alpha D^\mu A_\beta\right) S^{\alpha\beta}$$
$$+ \frac{1}{2g^2} e\, e^{-2(D-2)a\phi} \mathrm{tr}\left([A_\alpha, A_\beta][A_\gamma, A_\delta]\right) S^{\alpha\gamma} S^{\beta\delta}. \tag{6.23}$$

This Lagrangian is invariant up to total divergences under the D-dimensional general coordinate transformation, the ξ^α gauge transformation and the Yang–Mills gauge transformation (6.21). This Lagrangian also has the global symmetry $SL(d, \mathbb{R})$, which appeared in the dimensional reduction of the gravitational field. However, it does not have the global $\mathbb{R}^+$ symmetry unless the gauge group is Abelian.

The scalar fields A_α have a vacuum expectation value which minimize the potential term in (6.23). Since the potential is a square of the commutator of the matrix-valued scalar fields, the vacuum expectation values are represented by mutually commuting matrices in the Cartan subalgebra of the gauge group. Fluctuations of the scalar fields and the Yang–Mills fields around such a vacuum generically become massive due to the scalar field expectation values. However, fluctuations in the directions of the Cartan subalgebra commute with the vacuum expectation values and are massless. (For special vacuum expectation values, more massless fields appear exceptionally.) Therefore, when we are interested in massless fields in D dimensions, we only need to consider $\hat{D}$-dimensional Yang–Mills fields in the directions of the Cartan subalgebra. They are treated as Abelian gauge fields.

6.2.3 Antisymmetric Tensor Field

Let us consider a dimensional reduction of a second rank antisymmetric tensor field $\hat{B}_{MN}$ coupled to the gravitational field. Antisymmetric tensor fields of higher rank can be treated in a similar way. The $\hat{D}$-dimensional Lagrangian is

$$\mathscr{L} = -\frac{1}{12} \hat{e} \hat{F}_{MNP} \hat{F}^{MNP}, \tag{6.24}$$

where the field strength is $\hat{F}_{MNP} = 3\partial_{[M}\hat{B}_{NP]}$. The field strength and the Lagrangian are invariant under the gauge transformations $\delta\hat{B}_{MN} = 2\partial_{[M}\hat{\zeta}_{N]}$.

We assume that the field $\hat{B}_{MN}$ is independent of the compactified coordinates y^α. As in (6.20), (6.22) we define the D-dimensional fields

$$B_{\mu\nu} \equiv e^{2\sigma} e_\mu{}^a e_\nu{}^b \hat{B}_{ab}, \quad B_{\mu\alpha} \equiv e^\sigma e_\mu{}^a E_\alpha{}^i \hat{B}_{ai}, \quad B_{\alpha\beta} \equiv E_\alpha{}^i E_\beta{}^j \hat{B}_{ij} \qquad (6.25)$$

and the field strengths

$$\begin{aligned}
F_{\mu\nu\rho} &\equiv e^{3\sigma} e_\mu{}^a e_\nu{}^b e_\rho{}^c \hat{F}_{abc} = F_{\mu\nu\rho}^{(0)} - 3F_{[\mu\nu}{}^\alpha B_{\rho]\alpha}, \\
F_{\mu\nu\alpha} &\equiv e^{2\sigma} e_\mu{}^a e_\nu{}^b E_\alpha{}^i \hat{F}_{abi} = F_{\mu\nu\alpha}^{(0)} + F_{\mu\nu}{}^\beta B_{\beta\alpha},
\end{aligned} \qquad (6.26)$$

from the $\hat{D}$-dimensional ones with local Lorentz indices $\hat{B}_{AB}$, $\hat{F}_{ABC}$, where $F_{\mu\nu\rho}^{(0)} = 3\partial_{[\mu}B_{\nu\rho]}$, $F_{\mu\nu\alpha}^{(0)} = 2\partial_{[\mu}B_{\nu]\alpha}$. They are invariant under the ξ^α transformation. The Lagrangian (6.24) then becomes

$$\begin{aligned}
\mathcal{L} = &-\frac{1}{12} e\, e^{2da\phi} F_{\mu\nu\rho} F^{\mu\nu\rho} - \frac{1}{4} e\, e^{-(D-d-2)a\phi} F_{\mu\nu\alpha} F^{\mu\nu}{}_\beta S^{\alpha\beta} \\
&- \frac{1}{4} e\, e^{-2(D-2)a\phi} \partial_\mu B_{\alpha\beta} \partial^\mu B_{\gamma\delta} S^{\alpha\gamma} S^{\beta\delta}.
\end{aligned} \qquad (6.27)$$

This is a theory of an antisymmetric tensor field $B_{\mu\nu}$, d vector fields $B_{\mu\alpha}$ and $\frac{1}{2}d(d-1)$ scalar fields $B_{\alpha\beta}$ in D dimensions.

This Lagrangian is invariant up to total divergences under the D-dimensional general coordinate transformation and the ξ^α gauge transformation. It is also invariant under the gauge transformation derived from the $\hat{D}$-dimensional gauge transformation $\delta\hat{B}_{MN} = 3\partial_{[M}\hat{\zeta}_{N]}$ with a parameter independent of y^α. They are given by

$$\delta B_{\mu\nu} = 2\partial_{[\mu}\zeta_{\nu]} + F_{\mu\nu}{}^\alpha \zeta_\alpha, \quad \delta B_{\mu\alpha} = \partial_\mu \zeta_\alpha, \quad \delta B_{\alpha\beta} = 0, \qquad (6.28)$$

where the transformation parameters are related to $\hat{\zeta}_A = \hat{e}_A{}^M \hat{\zeta}_M$ as $\zeta_\mu = e^\sigma e_\mu{}^a \hat{\zeta}_a$, $\zeta_\alpha = E_\alpha{}^i \hat{\zeta}_i$. The parameter ζ_α represents $U(1)^d$ gauge transformations of the vector fields $B_{\mu\alpha}$. Note that $B_{\mu\nu}$ also transforms under this $U(1)^d$.

The Lagrangian (6.27) has the global $SL(d, \mathbb{R}) \times \mathbb{R}^+$ symmetry, which we discussed in Sect. 6.2.1. The $\mathbb{R}^+$ transformations are given by (6.17) and

$$\delta B_{\mu\nu} = -\frac{2d}{\hat{D}-2} c\, B_{\mu\nu}, \quad \delta B_{\mu\alpha} = \frac{D-d-2}{\hat{D}-2} c\, B_{\mu\alpha}, \quad \delta B_{\alpha\beta} = \frac{2D-4}{\hat{D}-2} c\, B_{\alpha\beta}. \qquad (6.29)$$

It is also invariant under the global transformations

$$\delta B_{\mu\nu} = 2A_{[\mu}{}^\alpha \zeta_{\nu]\alpha} + A_\mu{}^\alpha A_\nu{}^\beta \zeta_{\alpha\beta}, \quad \delta B_{\mu\alpha} = \zeta_{\mu\alpha} + \zeta_{\alpha\beta} A_\mu{}^\beta, \quad \delta B_{\alpha\beta} = \zeta_{\alpha\beta}, \qquad (6.30)$$

where $\zeta_{\mu\alpha}$ and $\zeta_{\alpha\beta} = -\zeta_{\beta\alpha}$ are constant parameters. These transformations are derived from the $\hat{D}$-dimensional gauge transformation $\delta\hat{B}_{MN} = 3\partial_{[M}\hat{\zeta}_{N]}$ with parameters of the form $\hat{\zeta}_\mu = -\zeta_{\mu\alpha}y^\alpha$, $\hat{\zeta}_\alpha = -\frac{1}{2}\zeta_{\alpha\beta}y^\beta$.

In the above discussion we have used the D-dimensional fields defined in (6.25), which are invariant under the ξ^α transformation. In some cases it is more convenient to use other definitions of D-dimensional fields. For instance, when one considers a dimensional reduction of supergravity, a global symmetry in D dimensions becomes manifest for a definition of the fields different from (6.25).

6.3 Dimensional Reductions of $D = 11$, $\mathcal{N} = 1$ Supergravity

In this section we consider dimensional reductions of $\hat{D} = 11$, $\mathcal{N} = 1$ supergravity discussed in Sect. 5.2 to $D = 11 - d$ dimensions. Since the number of supersymmetries is preserved by dimensional reductions, the D-dimensional theories are maximal supergravities. For simplicity we consider only the bosonic fields. The bosonic fields of the $\hat{D} = 11$ theory are a gravitational field $\hat{e}_M{}^A$ and a third rank antisymmetric tensor field $\hat{B}_{MNP}$.

From the gravitational field in 11 dimensions we obtain a gravitational field $e_\mu{}^a$, d vector fields $A_\mu{}^\alpha$ and $\frac{1}{2}d(d+1)$ scalar fields ϕ, $S_{\alpha\beta}$ in D dimensions as we saw in Sect. 6.2.1. The D-dimensional Lagrangian is (6.12). From the antisymmetric tensor field in 11 dimensions we obtain a third rank antisymmetric tensor field $B_{\mu\nu\rho}$, d second rank antisymmetric tensor fields $B_{\mu\nu\alpha}$, $\frac{1}{2}d(d-1)$ vector fields $B_{\mu\alpha\beta}$ and $\frac{1}{6}d(d-1)(d-2)$ scalar fields $B_{\alpha\beta\gamma}$. The scalar fields are chosen as $B_{\alpha\beta\gamma} = \hat{B}_{\alpha\beta\gamma}$. We define the field strengths of the tensor and vector fields in D dimensions from $\hat{F}_{ABCD}$ with local Lorentz indices as in (6.26) so that they are invariant under the ξ^α transformation. We then find the D-dimensional Lagrangian as

$$\mathcal{L} = eR - \frac{1}{2}e\partial_\mu\phi\partial^\mu\phi + \frac{1}{4}e\partial_\mu S^{\alpha\beta}\partial^\mu S_{\alpha\beta} - \frac{1}{4}e\,e^{9a\phi}S_{\alpha\beta}F_{\mu\nu}{}^\alpha F^{\mu\nu\beta}$$
$$- \frac{1}{12}e\,e^{3(d-9)a\phi}S^{\alpha\delta}S^{\beta\varepsilon}S^{\gamma\eta}\partial_\mu B_{\alpha\beta\gamma}\partial^\mu B_{\delta\varepsilon\eta}$$
$$- \frac{1}{8}e\,e^{3(d-6)a\phi}S^{\alpha\gamma}S^{\beta\delta}F_{\mu\nu\alpha\beta}F^{\mu\nu}{}_{\gamma\delta} - \frac{1}{48}e\,e^{3da\phi}F_{\mu\nu\rho\sigma}F^{\mu\nu\rho\sigma}$$
$$- \frac{1}{12}e\,e^{3(d-3)a\phi}S^{\alpha\beta}F_{\mu\nu\rho\alpha}F^{\mu\nu\rho}{}_\beta + \mathcal{L}_{\text{CS}}, \tag{6.31}$$

where $\mathcal{L}_{\text{CS}}$ is the third term (the Chern–Simons term) in (5.1) and the constant a is given in (6.11) with $D = 11 - d$.

The D-dimensional theory has the symmetries under the general coordinate transformation and the ξ^α gauge transformation. (When the fermionic fields are included, it also has the symmetries under the local Lorentz transformation and the local supertransformation.) It also has the symmetries under the gauge transformations of the antisymmetric tensor and vector fields derived from those of $\hat{B}_{MNP}$ similar to (6.28). The D-dimensional theory has the global symmetries under $SL(d, \mathbb{R}) \times \mathbb{R}^+$ discussed in Sect. 6.2.1 and under the transformation of the antisymmetric tensor fields similar to (6.30). Actually, the theories for $D \leq 8$ have larger global symmetries. These enhanced symmetries are the groups G listed in Table 4.1. They are not manifest in the Lagrangian (6.31) but can be seen by rewriting it in an appropriate way. In the following we discuss these global symmetries for each D (For more detailed discussion of the global symmetries in the dimensional reductions, see [4, 5]).

6.3.1 $D = 10$ Theory

For $D = 10$ we obtain $\mathcal{N} = (1, 1)$ supergravity discussed in Sect. 5.3. The bosonic fields in 10 dimensions are

$$e_\mu{}^a, \quad \phi, \quad A_\mu, \quad B_{\mu\nu}, \quad B_{\mu\nu\rho}, \tag{6.32}$$

where we have suppressed the indices $\alpha, \beta, \ldots = 10$. By appropriate field redefinitions the Lagrangian (6.31) reduces to the bosonic part of the Lagrangian (5.7). Including the fermionic fields the complete Lagrangian of $D = 10, \mathcal{N} = (1, 1)$ supergravity can be obtained by the dimensional reduction. The global symmetry is $G = \mathbb{R}^+$ as we saw in Sect. 5.3.

This relation between $D = 11$ and $D = 10$ supergravities supports a conjectured relation between M theory and type IIA superstring theory [18, 19] at the level of low energy effective theories. As we remarked after (6.12), the size of the compactified direction is $R = e^{\frac{1}{2}(D-2)a\phi} = e^{\frac{2}{3}\phi}$. Since the scalar field factor e^ϕ represents a string coupling constant g, we find a relation $R \sim g^{\frac{2}{3}}$. Therefore, for a weak coupling $g \ll 1$ the theory is described by type IIA superstring theory in 10 dimensions with small R, while for a strong coupling $g \gg 1$ it is described by M theory in 11 dimensions with large R.

6.3.2 $D = 9$ Theory

For $D = 9$ we obtain $\mathcal{N} = 2$ supergravity [1]. The bosonic fields are

$$e_\mu{}^a, \quad \phi, \quad S_{\alpha\beta}, \quad A_\mu{}^\alpha, \quad B_{\mu\nu\alpha}, \quad B_{\mu\nu\rho}, \tag{6.33}$$

where $\alpha, \beta, \ldots = 1, 2$. The Lagrangian (6.31) has the global symmetry $G = SL(2, \mathbb{R}) \times \mathbb{R}^+$. The kinetic terms of the scalar fields in the Lagrangian are

$$\mathcal{L} = -\frac{1}{2}e\partial_\mu\phi\partial^\mu\phi + \frac{1}{4}e\partial_\mu S^{\alpha\beta}\partial^\mu S_{\alpha\beta}. \tag{6.34}$$

The scalar fields $S_{\alpha\beta}$ and ϕ are described by the $SL(2,\mathbb{R})/SO(2) \times \mathbb{R}^+$ non-linear sigma model. The vector fields $A_\mu{}^\alpha$ and the antisymmetric tensor field $B_{\mu\nu\alpha}$ belong to the representation **2** of $SL(2,\mathbb{R})$.

6.3.3 $D = 8$ Theory

For $D = 8$ we obtain $\mathcal{N} = 2$ supergravity [13]. The bosonic fields are

$$e_\mu{}^a, \quad \phi, \quad S_{\alpha\beta}, \quad B, \quad A_\mu{}^\alpha, \quad B_{\mu\alpha\beta}, \quad B_{\mu\nu\alpha}, \quad B_{\mu\nu\rho}, \tag{6.35}$$

where the scalar field B is a component of the antisymmetric tensor field in 11 dimensions $\hat{B}_{\alpha\beta\gamma} = \varepsilon_{\alpha\beta\gamma}B$. The Lagrangian (6.31) has the global symmetries under the $SL(3,\mathbb{R}) \times \mathbb{R}^+$ transformation and constant shifts of B similar to (6.30). Actually, the theory has the larger symmetry $G = SL(3,\mathbb{R}) \times SL(2,\mathbb{R})$. The kinetic terms of the scalar fields are

$$\mathscr{L} = -\frac{1}{2}e\left(\partial_\mu\phi\partial^\mu\phi + e^{-2\phi}\partial_\mu B\partial^\mu B\right) + \frac{1}{4}e\partial_\mu S^{\alpha\beta}\partial^\mu S_{\alpha\beta}. \tag{6.36}$$

The last term represents the $SL(3,\mathbb{R})/SO(3)$ non-linear sigma model. The other two terms can be rewritten as follows. We introduce a 2×2 matrix-valued field

$$v' = \begin{pmatrix} e^{\frac{1}{2}\phi} & e^{-\frac{1}{2}\phi}B \\ 0 & e^{-\frac{1}{2}\phi} \end{pmatrix}, \quad \det v' = 1, \tag{6.37}$$

which is a representative of the coset $SL(2,\mathbb{R})/SO(2)$. Following Sect. 4.1 we define $v'^{-1}\partial_\mu v' = P'_\mu + Q'_\mu$ $(P'^T_\mu = P'_\mu, \ Q'^T_\mu = -Q'_\mu)$. Then, the kinetic terms of ϕ and B in (6.36) can be written as $-\mathrm{tr}\left(P'_\mu P'^\mu\right)$. Thus, we see that these two scalar fields are described by the $SL(2,\mathbb{R})/SO(2)$ non-linear sigma model.

The vector fields $(A_\mu{}^\alpha, B_\mu{}^\alpha)$, where $B_{\mu\beta\gamma} = \varepsilon_{\beta\gamma\alpha}B_\mu{}^\alpha$, belong to the representation $(\mathbf{3},\mathbf{2})$ of $G = SL(3,\mathbb{R}) \times SL(2,\mathbb{R})$. The second rank antisymmetric tensor field $B_{\mu\nu\alpha}$ belongs to the representation $(\mathbf{3},\mathbf{1})$ of G. The third rank antisymmetric tensor field $B_{\mu\nu\rho}$ does not transform under $SL(3,\mathbb{R})$ in G. $SL(2,\mathbb{R})$ acts on $B_{\mu\nu\rho}$ as the duality transformation as discussed in Sect. 4.2. Its field strength $F_{\mu\nu\rho\sigma}$ and $G_{\mu\nu\rho\sigma}$ defined by (4.23) together belong to the representation **2** of $SL(2,\mathbb{R})$. Thus, $SL(3,\mathbb{R})$ is a symmetry of the Lagrangian but $SL(2,\mathbb{R})$ is a symmetry of the field equations since it contains the duality transformation.

6.3.4 $D = 7$ Theory

For $D = 7$ we obtain $\mathcal{N} = 4$ supergravity [16]. The bosonic fields are

$$e_\mu{}^a, \quad \phi, \quad S_{\alpha\beta}, \quad B^\alpha, \quad A_\mu{}^\alpha, \quad B_{\mu\alpha\beta}, \quad B_{\mu\nu\alpha}, \quad B_{\mu\nu\rho}, \tag{6.38}$$

where the scalar fields B^α are components of the antisymmetric tensor field in 11 dimensions $\hat{B}_{\alpha\beta\gamma} = \varepsilon_{\alpha\beta\gamma\delta}B^\delta$. The Lagrangian (6.31) for these fields has the global symmetry $G = SL(5, \mathbb{R})$ larger than $SL(4, \mathbb{R}) \times \mathbb{R}^+$. The kinetic terms of the scalar fields are

$$\mathcal{L} = -\frac{1}{2}e\partial_\mu\phi\partial^\mu\phi + \frac{1}{4}e\partial_\mu S^{\alpha\beta}\partial^\mu S_{\alpha\beta} - \frac{1}{2}ee^{-\frac{5}{\sqrt{10}}\phi}S_{\alpha\beta}\partial_\mu B^\alpha\partial^\mu B^\beta. \qquad (6.39)$$

We can introduce a 5×5 matrix-valued scalar field

$$V = \begin{pmatrix} e^{\frac{2}{\sqrt{10}}\phi} & e^{-\frac{1}{2\sqrt{10}}\phi}B^\beta v_\beta{}^i \\ 0 & e^{-\frac{1}{2\sqrt{10}}\phi}v_\alpha{}^i \end{pmatrix}, \qquad \det V = 1, \qquad (6.40)$$

which is a representative of the coset $SL(5, \mathbb{R})/SO(5)$. As in the $D = 8$ case, (6.39) can be written as $-\mathrm{tr}(P_\mu P^\mu)$, where $V^{-1}\partial_\mu V = P_\mu + Q_\mu$ ($P_\mu^T = P_\mu$, $Q_\mu^T = -Q_\mu$). Therefore, the scalar fields are described by the $SL(5, \mathbb{R})/SO(5)$ non-linear sigma model. The vector fields $(A_\mu{}^\alpha, B_{\mu\alpha\beta})$ belong to the representation **10** of $SL(5, \mathbb{R})$. The third rank antisymmetric tensor field $B_{\mu\nu\rho}$ can be replaced by a dual field $\tilde{B}_{\mu\nu}$ as in Sect. 1.4. Then, the second rank antisymmetric tensor fields $(\tilde{B}_{\mu\nu}, B_{\mu\nu\alpha})$ belong to the representation **5** of $SL(5, \mathbb{R})$.

6.3.5 $D = 6$ Theory

For $D = 6$ we obtain $\mathcal{N} = (4, 4)$ supergravity [17]. The bosonic fields are

$$e_\mu{}^a, \quad \phi, \quad S_{\alpha\beta}, \quad B^{\alpha\beta}, \quad A_\mu{}^\alpha, \quad B_{\mu\alpha\beta}, \quad B_{\mu\nu\alpha}, \quad B_{\mu\nu\rho}, \qquad (6.41)$$

where the scalar fields $B^{\alpha\beta}$ are components of the antisymmetric tensor field in 11 dimensions $\hat{B}_{\alpha\beta\gamma} = \frac{1}{2}\varepsilon_{\alpha\beta\gamma\delta\varepsilon}B^{\delta\varepsilon}$. The field equations have the global symmetry $G = SO(5, 5)$ larger than $SL(5, \mathbb{R}) \times \mathbb{R}^+$. The kinetic terms of the scalar fields are

$$\mathcal{L} = -\frac{1}{2}e\partial_\mu\phi\partial^\mu\phi + \frac{1}{4}e\partial_\mu S^{\alpha\beta}\partial^\mu S_{\alpha\beta} - \frac{1}{4}ee^{-\frac{4}{\sqrt{10}}\phi}S_{\alpha\gamma}S_{\beta\delta}\partial_\mu B^{\alpha\beta}\partial^\mu B^{\gamma\delta}.$$
$$(6.42)$$

We can introduce a 10×10 matrix-valued scalar field

$$V = \begin{pmatrix} V^\alpha{}_i & V^{\alpha i} \\ V_{\alpha i} & V_\alpha{}^i \end{pmatrix} = \begin{pmatrix} e^{\frac{1}{\sqrt{10}}\phi}v_i{}^\alpha & e^{-\frac{1}{\sqrt{10}}\phi}B^{\alpha\beta}v_\beta{}^i \\ 0 & e^{-\frac{1}{\sqrt{10}}\phi}v_\alpha{}^i \end{pmatrix}, \qquad (6.43)$$

which is a representative of the coset $SO(5, 5)/[SO(5) \times SO(5)]$. In fact, V satisfies $V^T \Omega V = \Omega$ for Ω in (4.32) with $\varepsilon = +1$. The kinetic terms (6.42) can be written as $-\frac{1}{8}\text{tr}(P_\mu P^\mu)$, where $V^{-1}\partial_\mu V = P_\mu + Q_\mu$ ($P_\mu^T = P_\mu$, $Q_\mu^T = -Q_\mu$). Therefore, the scalar fields are described by the $SO(5, 5)/[SO(5) \times SO(5)]$ non-linear sigma model. The third rank antisymmetric tensor field $B_{\mu\nu\rho}$ can be replaced by a dual vector field $\tilde{B}_\mu$. Then, the vector fields $(A_\mu{}^\alpha, B_{\mu\alpha\beta}, \tilde{B}_\mu)$ belong to the representation **16** of $SO(5, 5)$. $SO(5, 5)$ acts on the antisymmetric tensor fields $B_{\mu\nu\alpha}$ as the duality transformation. The field strengths $F_{\mu\nu\rho\alpha}$ and $G_{\mu\nu\rho}{}^\alpha$ defined by (4.23) together belong to the representation **10** of $SO(5, 5)$.

6.3.6 $D = 5$ Theory

For $D = 5$ we obtain $\mathcal{N} = 8$ supergravity [2, 6]. The bosonic fields are

$$e_\mu{}^a, \quad \phi, \quad S_{\alpha\beta}, \quad B_{\alpha\beta\gamma}, \quad A_\mu{}^\alpha, \quad B_{\mu\alpha\beta}, \quad B_{\mu\nu\alpha}, \quad B_{\mu\nu\rho}. \tag{6.44}$$

$B_{\mu\nu\rho}$ and $B_{\mu\nu\alpha}$ can be replaced by dual fields $\tilde{B}$ and $\tilde{B}_\mu{}^\alpha$, respectively. The Lagrangian has the global symmetry $G = E_{6(+6)}$ larger than $SL(6, \mathbb{R}) \times \mathbb{R}^+$. The kinetic terms of the scalar fields are

$$\mathcal{L} = -\frac{1}{2}e\partial_\mu\phi\partial^\mu\phi - \frac{1}{2}e\,e^{-2\phi}\left(\partial_\mu\tilde{B} - \frac{1}{2}(B\partial_\mu B)\right)\left(\partial^\mu\tilde{B} - \frac{1}{2}(B\partial^\mu B)\right)$$

$$+ \frac{1}{4}e\partial_\mu S^{\alpha\beta}\partial^\mu S_{\alpha\beta} - \frac{1}{12}e\,e^{-\phi}S^{\alpha\delta}S^{\beta\varepsilon}S^{\gamma\eta}\partial_\mu B_{\alpha\beta\gamma}\partial^\mu B_{\delta\varepsilon\eta}, \tag{6.45}$$

where $(B\partial_\mu B) = \frac{1}{6}\varepsilon^{\alpha\beta\gamma\delta\varepsilon\eta}B_{\alpha\beta\gamma}\partial_\mu B_{\delta\varepsilon\eta}$. The term $(B\partial_\mu B)$ appears when $B_{\mu\nu\rho}$ is replaced by the dual field $\tilde{B}$. We can introduce a 27×27 matrix-valued scalar field

$$V = \begin{pmatrix} e^{\frac{1}{2}\phi}v_\alpha{}^i & B_{\alpha ij} & -e^{-\frac{1}{2}\phi}\left(\tilde{B}v_\alpha{}^i + \frac{1}{4}B_{\alpha\gamma\delta}\tilde{B}^{\gamma\delta i}\right) \\ 0 & 2v_{[i}{}^\alpha v_{j]}{}^\beta & -e^{-\frac{1}{2}\phi}\tilde{B}^{\alpha\beta i} \\ 0 & 0 & e^{-\frac{1}{2}\phi}v_\alpha{}^i \end{pmatrix}, \tag{6.46}$$

where $B_{\alpha ij} = v_i{}^\gamma v_j{}^\delta B_{\alpha\gamma\delta}$, $\tilde{B}^{\alpha\beta i} = \tilde{B}^{\alpha\beta\gamma}v_\gamma{}^i$. The indices of the rows and the columns of this matrix are $\alpha, [\alpha\beta], \alpha$ and $i, [ij], i$, respectively. This matrix is a representative of the coset $E_{6(+6)}/USp(8)$. $E_{6(+6)}$ is a non-compact Lie group of dimension 78 and $USp(8)$ is its maximal compact subgroup (For more details of $E_{6(+6)}$, see [2, 6, 9]). The kinetic terms (6.45) can be written as $-\frac{1}{6}\text{tr}\left(P_\mu P^\mu\right)$, where $V^{-1}\partial_\mu V = P_\mu + Q_\mu$ ($P_\mu^T = P_\mu$, $Q_\mu^T = -Q_\mu$). Therefore, the scalar fields are described by the $E_{6(+6)}/USp(8)$ non-linear sigma model. The vector fields $(A_\mu{}^\alpha, B_{\mu\alpha\beta}, \tilde{B}_\mu{}^\alpha)$ belong to the representation **27** of $E_{6(+6)}$.

6.3.7 $D = 4$ Theory

For $D = 4$ we obtain $\mathcal{N} = 8$ supergravity discussed in Sect. 4.3. The bosonic fields are

$$e_\mu{}^a, \quad \phi, \quad S_{\alpha\beta}, \quad B_{\alpha\beta\gamma}, \quad A_\mu{}^\alpha, \quad B_{\mu\alpha\beta}, \quad B_{\mu\nu\alpha}, \quad B_{\mu\nu\rho}. \tag{6.47}$$

$B_{\mu\nu\alpha}$ can be replaced by a dual scalar field $\tilde{B}^\alpha$. $B_{\mu\nu\rho}$ does not have physical degrees of freedom since the solution of its field equations is $F_{\mu\nu\rho\sigma} = e\varepsilon_{\mu\nu\rho\sigma} \times$ (constant). The field equations have the global symmetry $G = E_{7(+7)}$ larger than $SL(7, \mathbb{R}) \times \mathbb{R}^+$. The kinetic terms of the scalar fields are

$$\mathcal{L} = -\frac{1}{2}e\partial_\mu\phi\partial^\mu\phi + \frac{1}{4}e\partial_\mu S^{\alpha\beta}\partial^\mu S_{\alpha\beta} - \frac{1}{12}e\,e^{-\frac{2}{\sqrt{7}}\phi}S^{\alpha\delta}S^{\beta\varepsilon}S^{\gamma\eta}\partial_\mu B_{\alpha\beta\gamma}\partial^\mu B_{\delta\varepsilon\eta}$$
$$-\frac{1}{2}e\,e^{-\frac{4}{\sqrt{7}}\phi}S_{\alpha\beta}\left(\partial_\mu\tilde{B}^\alpha - \frac{1}{12}(B\partial_\mu B)^\alpha\right)\left(\partial^\mu\tilde{B}^\beta - \frac{1}{12}(B\partial^\mu B)^\beta\right), \tag{6.48}$$

where $(B\partial_\mu B)^\alpha = \frac{1}{6}\varepsilon^{\alpha\gamma_1\cdots\gamma_6}B_{\gamma_1\gamma_2\gamma_3}\partial_\mu B_{\gamma_4\gamma_5\gamma_6}$. We can define a 56×56 matrix-valued scalar field V by using the fields $\phi, S_{\alpha\beta}, B_{\alpha\beta\gamma}, \tilde{B}^\alpha$, which is a representative of the coset $E_{7(+7)}/SU(8)$ (For the explicit form of the matrix, see [3]). The kinetic terms (6.48) can be written as $-\frac{1}{12}\mathrm{tr}\left(P_\mu P^\mu\right)$, where $V^{-1}\partial_\mu V = P_\mu + Q_\mu$ ($P_\mu^T = P_\mu, Q_\mu^T = -Q_\mu$). Therefore, the scalar fields are described by the $E_{7(+7)}/SU(8)$ non-linear sigma model. $E_{7(+7)}$ acts on the vector fields $(A_\mu{}^\alpha, B_{\mu\alpha\beta})$ as the duality transformation. The field strengths $F_{\mu\nu}$ and $G_{\mu\nu}$ defined by (4.23) together belong to the representation **56** of $E_{7(+7)}$.

6.4 Dimensional Reductions of $D = 10$, $\mathcal{N} = (2, 0)$ Supergravity

$D = 10$, $\mathcal{N} = (2, 0)$ supergravity discussed in Sect. 5.4 cannot be derived by a dimensional reduction of $D = 11$ supergravity. Hence, its reductions to lower dimensions must be studied separately. In this section we consider its dimensional reduction to $D = 9$ [1]. As we will see below, the $D = 9$ theory becomes $\mathcal{N} = 2$ supergravity, which is the same as the $D = 9$ theory obtained by the reduction from $D = 11$. Therefore, reductions to $D < 9$ are the same as those from $D = 11$ supergravity.

The bosonic fields of $\hat{D} = 10$, $\mathcal{N} = (2, 0)$ supergravity are a gravitational field $\hat{e}_M{}^A$, two real scalar fields $\hat{\varphi}$, $\hat{C}$, two second rank antisymmetric tensor fields $\hat{B}_{MN(\alpha)}$ ($\alpha = 1, 2$)[1] and a fourth rank antisymmetric tensor field $\hat{B}_{MNPQ}$. By the dimensional reduction to $D = 9$ we find the bosonic fields in nine dimensions

---

[1] Here, $\alpha, \beta, \ldots$ are indices which distinguish two second rank antisymmetric tensor fields and are not indices of the compactified directions. They correspond to $i, j, \ldots$ in (5.42).

$$e_\mu{}^a, \quad \phi, \quad \varphi, \quad C, \quad A_\mu, \quad \overset{\prime}{B}_\mu{}^{(\alpha)}, \quad B_{\mu\nu(\alpha)}, \quad B_{\mu\nu\rho}, \quad B_{\mu\nu\rho\sigma}. \tag{6.49}$$

The relations of these fields to the 10-dimensional fields are (6.6) and $\hat{\varphi} = \varphi$, $\hat{C} = C, \hat{B}_{\mu 9(\alpha)} = \varepsilon_{\alpha\beta} B_\mu{}^{(\beta)}, \hat{B}_{\mu\nu(\alpha)} = B_{\mu\nu(\alpha)} + \cdots, \hat{B}_{\mu\nu 9} = B_{\mu\nu\rho} + \cdots, \hat{B}_{\mu\nu\rho\sigma} = B_{\mu\nu\rho\sigma} + \cdots$, where $\ldots$ denote terms which are products of lower rank fields. The field strengths of the vector and antisymmetric tensor fields are defined from local Lorentz components of the 10-dimensional field strengths as in (6.26).

The nine-dimensional Lagrangian derived from (5.42) is

$$\mathcal{L} = eR - \frac{1}{2}e\partial_\mu\phi\partial^\mu\phi + \frac{1}{4}\partial_\mu S'^{\alpha\beta}\partial^\mu S'_{\alpha\beta} - \frac{1}{4}e\, e^{\frac{4}{\sqrt{7}}\phi} F_{\mu\nu}F^{\mu\nu}$$
$$- \frac{1}{4}e\, e^{-\frac{3}{\sqrt{7}}\phi} S'_{\alpha\beta} F_{\mu\nu}{}^{(\alpha)} F^{\mu\nu(\beta)} - \frac{1}{12}e\, e^{\frac{1}{\sqrt{7}}\phi} S'^{\alpha\beta} F_{\mu\nu\rho(\alpha)} F^{\mu\nu\rho}{}_{(\beta)}$$
$$- \frac{1}{96}e\, e^{-\frac{2}{\sqrt{7}}\phi} F_{\mu\nu\rho\sigma} F^{\mu\nu\rho\sigma} - \frac{1}{480}e\, e^{\frac{2}{\sqrt{7}}\phi} F_{\mu\nu\rho\sigma\tau} F^{\mu\nu\rho\sigma\tau} + \mathcal{L}_{CS}, \tag{6.50}$$

where $\mathcal{L}_{CS}$ is the last term (Chern–Simons term) of (5.42). We have introduced a 2×2 matrix $S'_{\alpha\beta}$ made from the scalar fields φ and C

$$S'_{\alpha\beta} = \begin{pmatrix} e^{-\varphi} + e^\varphi C^2 & e^\varphi C \\ e^\varphi C & e^\varphi \end{pmatrix}, \qquad \det S' = 1 \tag{6.51}$$

and its inverse $S'^{\alpha\beta}$. The self-duality condition (5.29) reduces to

$$e\, e^{-\frac{2}{\sqrt{7}}\phi} F^{\mu_1\cdots\mu_4} = -\frac{1}{5!}\varepsilon^{\mu_1\cdots\mu_9} F_{\mu_5\cdots\mu_9}. \tag{6.52}$$

Since this condition implies that $B_{\mu\nu\rho\sigma}$ and $B_{\mu\nu\rho}$ are related up to gauge degrees of freedom, we can use only $B_{\mu\nu\rho}$ as an independent field. Differentiating both-hand sides of (6.52) we find

$$\partial_{\mu_1}(e\, e^{-\frac{2}{\sqrt{7}}\phi} F^{\mu_1\cdots\mu_4}) = (\text{terms independent of } B_{\mu\nu\rho\sigma}). \tag{6.53}$$

This is a Maxwell type field equation of $B_{\mu\nu\rho}$ and does not contain $B_{\mu\nu\rho\sigma}$. The field equations of other fields can be derived from the Lagrangian (6.50) and are independent of $B_{\mu\nu\rho\sigma}$ when (6.52) is used. All the field equations obtained in this way can be derived from a new Lagrangian, which is independent of $B_{\mu\nu\rho\sigma}$. It can be shown that the new Lagrangian has the same form as that of the nine-dimensional theory derived from the $\hat{D} = 11$ supergravity by the dimensional reduction.

Thus, both of $\hat{D} = 10$, $\mathcal{N} = (2, 0)$ supergravity and $\hat{D} = 11$, $\mathcal{N} = 1$ supergravity reduce to the same $D = 9$ supergravity. However, the origins of the nine-dimensional fields are different. For instance, the scalar fields $S_{\alpha\beta}$ came from the 11-dimensional gravitational field but the corresponding $S'_{\alpha\beta}$ came from the scalar fields $\hat{\varphi}$, $\hat{C}$ already present in $\hat{D} = 10$, $\mathcal{N} = (2, 0)$ supergravity. The origins of the global symmetry

$SL(2, \mathbb{R})$ in nine dimensions are also different. For the reduction from 11 dimensions it arises from the coordinate transformation in the compactified directions. For the reduction from 10 dimensions it is a symmetry that already exists in $\hat{D} = 10$, $\mathcal{N} = (2, 0)$ supergravity.

Since $D = 10$, $\mathcal{N} = (1, 1)$ supergravity is a dimensional reduction of $\hat{D} = 11$, $\mathcal{N} = 1$ supergravity, the nine-dimensional theory can be obtained also by a dimensional reduction of the $D = 10$ theory. Therefore, $\mathcal{N} = (1, 1)$ and $\mathcal{N} = (2, 0)$ supergravities in 10 dimensions give the same nine-dimensional theory. These two 10-dimensional theories are low energy effective theories of type IIA and type IIB superstring theories, respectively. It is known that the type IIA and type IIB superstring theories compactified by S^1 are equivalent theories related by the T duality [7, 8]. The above result shows the T duality at the level of low energy effective theories.

6.5 Dimensional Reductions of $D = 10$, $\mathcal{N} = (1, 0)$ Supergravity

Finally, let us consider dimensional reductions of $\hat{D} = 10$, $\mathcal{N} = (1, 0)$ supergravity [10]. In this case the super Yang–Mills multiplets can be coupled to the supergravity multiplet as matter multiplets. The bosonic fields in 10 dimensions are a gravitational field $\hat{e}_M{}^A$, a real scalar field $\hat{\phi}$, a second rank antisymmetric tensor field $\hat{B}_{MN}$ and Yang–Mills fields $\hat{A}_M^I$. As we saw in Sect. 6.2.2, vector fields and scalar fields appear in lower dimensions by the dimensional reduction of the Yang–Mills fields. For a generic vacuum expectation value of the scalar fields, only the fields in the Cartan subalgebra of the gauge group are massless. Hence, we consider only the fields in the Cartan subalgebra $\hat{A}_M^I$ ($I = 1, 2, \ldots, n$) in 10 dimensions, where n is the rank of the gauge group. These fields can be regarded as $U(1)^n$ Abelian gauge fields.

We reduce this theory to $D = 10 - d$ dimensions. The bosonic fields in D dimensions are a gravitational field $e_\mu{}^a$, $d^2 + dn + 1$ scalar fields ϕ, $G_{\alpha\beta}$, $B_{\alpha\beta}$, $a_\alpha{}^I$, $2d + n$ vector fields $A_\mu{}^\alpha$, $B_{\mu\alpha}$, $A_\mu{}^I$ and an antisymmetric tensor field $B_{\mu\nu}$. The scalar fields are related to the 10-dimensional fields as

$$G_{\alpha\beta} = e^{\frac{1}{2}\hat{\phi}} E_\alpha{}^i E_{\beta i}, \quad \hat{B}_{\alpha\beta} = B_{\alpha\beta}, \quad e^{\hat{\phi}} = e^{\sqrt{\frac{8}{D-2}}\phi} E^{\frac{4}{D-2}}, \quad \hat{A}_\alpha^I = a_\alpha{}^I. \qquad (6.54)$$

Note that in this section we use the above new definition of $G_{\alpha\beta}$ different from the previous one $G_{\alpha\beta} = E_\alpha{}^i E_{\beta i}$ in (6.8). The field strengths of the vector and antisymmetric tensor fields are defined from local Lorentz components of the 10-dimensional field strengths as in (6.22), (6.26). The D-dimensional Lagrangian derived from (5.49) is

$$\mathcal{L} = eR - \frac{1}{2}e\partial_\mu\phi\partial^\mu\phi + \frac{1}{4}e\partial_\mu G^{\alpha\beta}\partial^\mu G_{\alpha\beta} - \frac{1}{4}e\partial_\mu B_{\alpha\beta}\partial^\mu B_{\gamma\delta}G^{\alpha\gamma}G^{\beta\delta}$$

$$- \frac{1}{4}e\,e^{-\sqrt{\frac{2}{D-2}}\phi}F_{\mu\nu}{}^\alpha F^{\mu\nu\beta}G_{\alpha\beta} - \frac{1}{4}e\,e^{-\sqrt{\frac{2}{D-2}}\phi}F_{\mu\nu\alpha}F^{\mu\nu}{}_\beta G^{\alpha\beta}$$

$$- \frac{1}{12}e\,e^{-\sqrt{\frac{8}{D-2}}\phi}F_{\mu\nu\rho}F^{\mu\nu\rho} - \frac{1}{4}e\,e^{-\sqrt{\frac{2}{D-2}}\phi}F_{\mu\nu}{}^I F^{\mu\nu I} - \frac{1}{2}e\partial_\mu a_\alpha{}^I\partial^\mu a_\beta{}^I G^{\alpha\beta}.$$

$$(6.55)$$

When the fermionic fields are included, the D-dimensional theory has a half of the maximal supersymmetry and consists of a supergravity multiplet and gauge multiplets. To see this, the antisymmetric tensor field $B_{\mu\nu}$ must be replaced by an equivalent dual field (a scalar field or a vector field) for $D = 4, 5$.

Global Symmetry

The Lagrangian (6.55) has the global $SO(d, d+n) \times \mathbb{R}^+$ symmetry. To see this, we collect the $2d + n$ vector fields into

$$\mathcal{A}_\mu = \begin{pmatrix} A_\mu{}^\alpha \\ B_{\mu\alpha} \\ A_\mu{}^I \end{pmatrix}, \qquad \mathcal{F}_{\mu\nu} = 2\partial_{[\mu}\mathcal{A}_{\nu]}. \qquad (6.56)$$

We also define a $(2d + n) \times (2d + n)$ matrix-valued scalar field

$$\mathcal{M} = \begin{pmatrix} G^{-1} & -G^{-1}C & -G^{-1}a \\ -C^T G^{-1} & G + C^T G^{-1}C + aa^T & a + C^T G^{-1}a \\ -a^T G^{-1} & a^T + a^T G^{-1}C & 1 + a^T G^{-1}a \end{pmatrix}, \qquad (6.57)$$

where G, G^{-1}, a, C are matrices with components $G_{\alpha\beta}$, $G^{\alpha\beta}$, $a_\alpha{}^I$, $C_{\alpha\beta} = B_{\alpha\beta} + \frac{1}{2}a_\alpha{}^I a_\beta{}^I$, respectively. The matrix $\mathcal{M}$ satisfies

$$\mathcal{M}^T = \mathcal{M}, \qquad \mathcal{M}^T \eta \mathcal{M} = \eta, \qquad \eta = \begin{pmatrix} 0 & 1 & 0 \\ 1 & 0 & 0 \\ 0 & 0 & 1 \end{pmatrix}. \qquad (6.58)$$

Diagonalizing η we have d eigenvalues -1 and $d+n$ eigenvalues $+1$. Therefore, the second equation of (6.58) implies that $\mathcal{M}$ is an element of $SO(d, d+n)$. Since $\mathcal{M}$ is also symmetric, there are $d^2 + dn$ independent components. This number is equal to the number of the scalar fields in $\mathcal{M}$, and the matrix (6.57) is the most general form which satisfies (6.58). Using these fields the Lagrangian (6.55) can be written as

$$\mathcal{L} = eR - \frac{1}{2}e\partial_\mu\phi\partial^\mu\phi + \frac{1}{8}e\,\mathrm{tr}\left(\partial_\mu\mathcal{M}^{-1}\partial^\mu\mathcal{M}\right)$$

$$- \frac{1}{4}e\,e^{-\sqrt{\frac{2}{D-2}}\phi}\mathcal{F}_{\mu\nu}^T \mathcal{M}^{-1}\mathcal{F}^{\mu\nu} - \frac{1}{12}e\,e^{-\sqrt{\frac{8}{D-2}}\phi}F_{\mu\nu\rho}F^{\mu\nu\rho}, \qquad (6.59)$$

where the field strength of the antisymmetric tensor field is given by

$$F_{\mu\nu\rho} = 3\partial_{[\mu}A_{\nu\rho]} - 3\mathcal{A}^T_{[\mu}\eta\partial_\nu\mathcal{A}_{\rho]}. \tag{6.60}$$

This Lagrangian is manifestly invariant under the global $SO(d, d+n)$ transformation

$$\mathcal{A}_\mu \to g\mathcal{A}_\mu, \qquad \mathcal{M} \to g\mathcal{M}g^T \qquad (g^T\eta g = \eta). \tag{6.61}$$

It is also invariant under the global $\mathbb{R}^+$ transformation

$$\phi \to \phi + \sqrt{2(D-2)}\,\alpha, \quad \mathcal{A}_\mu \to e^\alpha \mathcal{A}_\mu, \quad B_{\mu\nu} \to e^{2\alpha}B_{\mu\nu}. \tag{6.62}$$

An infinitesimal $SO(d, d+n)$ transformation can be written as $g = 1 + t$, where

$$t = \begin{pmatrix} a & b & c \\ e & -a^T & f \\ -f^T & -c^T & d \end{pmatrix} \qquad (b^T = -b,\ d^T = -d,\ e^T = -e). \tag{6.63}$$

The traceless part of the parameter a corresponds to $SL(d, \mathbb{R})$ discussed in Sect. 6.2.1. The trace part of a is a $\mathbb{R}^+$ transformation, which is different from (6.62). (There are two $\mathbb{R}^+$ symmetries in D-dimensional theory. One of them is the symmetry already present in 10 dimensions as we saw in Sect. 5.5. The other appears as a consequence of the dimensional reduction as in (6.17), (6.29).) The parameters e, f corresponds to constant shifts of the scalar fields in (6.30). The parameter d is the $SO(n)$ rotation of the n gauge fields $\hat{A}^I_M$. Other parameters b, c are transformations which exchange different fields $\hat{g}_{MN}$, $\hat{A}_{MN}$, $\hat{A}^I_M$ in 10 dimensions, and their origins are not obvious by the dimensional reduction.

When a 10-dimensional theory contains a gravitational field, D-dimensional theory has the global symmetry $SL(d, \mathbb{R}) \times \mathbb{R}^+$. When the 10-dimensional theory also contains an antisymmetric tensor field $B_{\mu\nu}$, this symmetry is enlarged to $SO(d, d) \times \mathbb{R}^+$. When n vector fields are added, the symmetry is further enlarged to $SO(d, d+n) \times \mathbb{R}^+$. The $SO(d, d+n)$ transformations mix fields in the supergravity multiplet and those in the super Yang–Mills multiplets. Such symmetries mixing fields in different supermultiplets are known to exist in other supergravity theories.

When the gauge group of the 10-dimensional theory is $SO(32)$ or $E_8 \times E_8$, this theory is a low energy effective theory of heterotic string theory. Heterotic string theory compactified by a d-dimensional torus has a symmetry $SO(d, d + n, \mathbb{Z})$, a discrete subgroup of $SO(d, d + n)$, which we have seen here [11, 12].

$SO(d, d + n)/[SO(d) \times SO(d + n)]$ Non-Linear Sigma Model

The $d^2 + dn$ scalar fields in the matrix $\mathcal{M}$ can be described by the $SO(d, d + n)/[SO(d) \times SO(d+n)]$ non-linear sigma model as follows. We define a $(2d + n) \times (2d + n)$ matrix-valued scalar field

$$V = \begin{pmatrix} V^\alpha{}_i & V^{\alpha i} & V^{\alpha \hat{I}} \\ V_{\alpha i} & V_\alpha{}^i & V_\alpha{}^{\hat{I}} \\ V_{Ii} & V_I{}^i & V_I{}^{\hat{I}} \end{pmatrix} = \begin{pmatrix} E_i'^\alpha & 0 & 0 \\ -E_i'^\beta C_{\beta\alpha} & E_\alpha'^{i} & a_\alpha{}^{\hat{I}} \\ -E_i'^\beta a_\beta{}^I & 0 & \delta_I^{\hat{I}} \end{pmatrix}, \tag{6.64}$$

where $E_\alpha'^{i} = e^{\frac{1}{4}\hat{\phi}} E_\alpha{}^i$, $E_i'^\alpha = e^{-\frac{1}{4}\hat{\phi}} E_i{}^\alpha$, and $\hat{I}, \hat{J}, \ldots = 1, 2, \ldots, n$. This V satisfies $V^T \eta V = \eta$ and is an element of $SO(d, d+n)$. Actually, V is a representative of the coset $SO(d, d+n)/[SO(d) \times SO(d+n)]$. The matrix $\mathscr{M}$ in (6.57) can be written as $\mathscr{M} = VV^T$. Following Sect. 4.1 we define $V^{-1}\partial_\mu V = P_\mu + Q_\mu$, where Q_μ is a part in the Lie algebra of $SO(d) \times SO(d+n)$ and P_μ is a part in its orthogonal complement. Explicitly, we find

$$P_\mu = \frac{1}{2} \begin{pmatrix} -X_{\mu ij} & -Y_{\mu ij} & -Z_{\mu i\hat{j}} \\ Y_{\mu ij} & X_{\mu ij} & Z_{\mu i\hat{j}} \\ -Z_{\mu j\hat{i}} & Z_{\mu j\hat{i}} & 0 \end{pmatrix}, \tag{6.65}$$

where $X_{\mu ij} = E_i'^\alpha \partial_\mu E_{\alpha j}' + E_j'^\alpha \partial_\mu E_{\alpha i}'$, $Y_{\mu ij} = E_i'^\alpha E_j'^\beta \left(\partial_\mu B_{\alpha\beta} + a_{[\alpha}{}^I \partial_\mu a_{\beta]}{}^I\right)$, $Z_{\mu i\hat{j}} = E_i'^\alpha \partial_\mu a_{\alpha\hat{j}}$. The kinetic term of the scalar fields can be written in the form (4.5) as

$$\frac{1}{8} e \operatorname{tr}\left(\partial_\mu \mathscr{M}^{-1}\partial^\mu \mathscr{M}\right) = -\frac{1}{2} e \operatorname{tr}\left(P_\mu P^\mu\right). \tag{6.66}$$

$D = 4$ Theory

For $D = 4$ the antisymmetric tensor field $B_{\mu\nu}$ can be replaced by a dual scalar field $\tilde{B}$. Then, we have a theory of an $\mathcal{N} = 4$ supergravity multiplet coupled to $n + 6$ gauge multiplets. As in Sect. 1.4, we can obtain a Lagrangian for the dual field $\tilde{B}$. Then, we find the bosonic Lagrangian as

$$\mathscr{L} = eR - \frac{1}{2} e \left(\partial_\mu \phi \partial^\mu \phi + e^{2\phi} \partial_\mu \tilde{B} \partial^\mu \tilde{B}\right) + \frac{1}{8} e \operatorname{tr}\left(\partial_\mu \mathscr{M}^{-1}\partial^\mu \mathscr{M}\right)$$
$$- \frac{1}{4} e\, e^{-\phi} \mathscr{F}_{\mu\nu}^T \mathscr{M}^{-1} \mathscr{F}^{\mu\nu} + \frac{1}{8} \varepsilon^{\mu\nu\rho\sigma} \tilde{B} \mathscr{F}_{\mu\nu}^T \eta \mathscr{F}_{\rho\sigma}. \tag{6.67}$$

The kinetic terms of ϕ and $\tilde{B}$ have the same form as (4.13), and these scalar fields can be described by the $SL(2, \mathbb{R})/SO(2)$ non-linear sigma model. The field equations of this theory has a global $SO(6, n+6) \times SL(2, \mathbb{R})$ symmetry larger than $SO(6, n+6) \times \mathbb{R}^+$ [15]. $SL(2, \mathbb{R})$ acts on the vector fields as the duality transformation discussed in Sect. 4.2. The field strengths of the vector fields $\mathscr{F}_{\mu\nu}$ and $G_{\mu\nu}$ defined in (4.23) belong to the representation $\mathbf{2}$ of $SL(2, \mathbb{R})$.

References

1. E. Bergshoeff, C.M. Hull, T. Ortín, Duality in the type II superstring effective action. Nucl. Phys. **B451**, 547 (1995). [hep-th/9504081]
2. E. Cremmer, Supergravities in 5 dimensions, in *Superspace and Supergravity*, ed. by S.W. Hawking, M. Roček (Cambridge University Press, Cambridge, 1981)
3. E. Cremmer, B. Julia, The SO(8) supergravity. Nucl. Phys. **B159**, 141 (1979)
4. E. Cremmer, B. Julia, H. Lü, C.N. Pope, Dualization of dualities. Nucl. Phys. **B523**, 73 (1998). [hep-th/9710119]
5. E. Cremmer, B. Julia, H. Lü, C.N. Pope, Dualization of dualities II: twisted self-duality of doubled fields and superdualities. Nucl. Phys. **B535**, 242 (1998). [hep-th/9806106]
6. E. Cremmer, J. Scherk, J.H. Schwarz, Spontaneously broken $\mathcal{N} = 8$ supergravity. Phys. Lett. **B84**, 83 (1979)
7. J. Dai, R.G. Leigh, J. Polchinski, New connections between string theories. Mod. Phys. Lett. **A4**, 2073 (1989)
8. M. Dine, P.Y. Huet, N. Seiberg, Large and small radius in string theory. Nucl. Phys. **B322**, 301 (1989)
9. R. Gilmore, *Lie Groups, Lie Algebras, and Some of Their Applications* (Wiley, New York, 1974)
10. J. Maharana, J.H. Schwarz, Noncompact symmetries in string theory. Nucl. Phys. **B390**, 3 (1993). [hep-th/9207016]
11. K.S. Narain, New heterotic string theories in uncompactified dimensions <10. Phys.Lett **B169**, 41 (1986)
12. K.S. Narain, M.H. Sarmadi, E. Witten, A note on toroidal compactification of heterotic string theory. Nucl. Phys. **B279**, 369 (1987)
13. A. Salam, E. Sezgin, $D = 8$ supergravity. Nucl. Phys. **B258**, 284 (1985)
14. J. Scherk, J.H. Schwarz, How to get masses from extra dimensions. Nucl. Phys. **B153**, 61 (1979)
15. A. Sen, Strong–weak coupling duality in four-dimensional string theory. Int. J. Mod. Phys. **A9**, 3707 (1994). [hep-th/9402002]
16. E. Sezgin, A. Salam, Maximal extended supergravity theory in seven-dimensions. Phys. Lett. **B118**, 359 (1982)
17. Y. Tanii, $\mathcal{N} = 8$ supergravity in six-dimensions. Phys. Lett. **B145**, 197 (1984)
18. P.K. Townsend, The eleven-dimensional supermembrane revisited. Phys. Lett. **B350**, 184 (1995). [hep-th/9501068]
19. E. Witten, String theory dynamics in various dimensions. Nucl. Phys. **B443**, 85 (1995). [hep-th/9503124]

Chapter 7
Gauged Supergravities

7.1 Gauged Supergravities and Massive Supergravities

In Poincaré supergravities discussed in Chaps. 2 and 5 there are no minimal couplings
of the vector fields to other fields. This means that the fields in the theories do not have
charges for the vector fields. In certain cases one can construct a theory which has
minimal couplings of the vector fields by a deformation of a Poincaré supergravity.
In those cases a gauge coupling constant g is introduced into the theory as a new
parameter. Such a theory is called the gauged supergravity. $D = 4$, $\mathcal{N} = 2$ AdS
supergravity discussed in Sect. 2.8 is an example of gauged supergravities. The gauge
groups of the vector fields in Poincaré supergravities are Abelian (a direct product of
$U(1)$'s), but the gauge groups of gauged supergravities are non-Abelian in general.

Another feature of gauged supergravities is that the Lagrangian contains Yukawa
couplings between the fermionic fields and the scalar fields proportional to g and
a scalar field potential proportional to g^2. (For theories without scalar fields, the
Lagrangian contains mass terms of the fermionic fields and a cosmological term.)
When the scalar field potential has a non-vanishing value for background scalar
fields, it plays a role of a cosmological term. Similarly, the Yukawa couplings for the
background scalar fields become mass terms of the fermionic fields. Some theories
have Yukawa couplings (or mass terms of the fermionic fields) and a scalar field
potential term (or a cosmological term) but do not contain minimal couplings to the
vector fields. $D = 4$, $\mathcal{N} = 1$ AdS supergravity discussed in Sect. 2.5 is an example.
Such a theory is called the massive supergravity. Since gauged supergravities and
massive supergravities have a cosmological term or a scalar potential term, they
sometimes have anti de Sitter (AdS) spacetime as a solution of the field equations.
Such a theory is also called the AdS supergravity as we already mentioned in Sect. 2.5.

Our discussion on gauged supergravities here is at an introductory level. There is
a systematic method to construct gauged supergravities using the embedding tensor.
For details of this method see [17] and references therein.

Y. Tanii, *Introduction to Supergravity*, SpringerBriefs in Mathematical Physics,   111
DOI: 10.1007/978-4-431-54828-7_7, © The Author(s) 2014

7.2 $D = 4$, $\mathcal{N} = 8$ Gauged Supergravity

As an example of gauged supergravities containing scalar fields let us consider a gauging of $D = 4$, $\mathcal{N} = 8$ supergravity discussed in Sect. 4.3 [3]. We gauge a subgroup $SO(8)$ of the global $E_{7(+7)}$ symmetry in the ungauged theory. As we saw in Sect. 4.3, $E_{7(+7)}$ is not a symmetry of the Lagrangian but is a symmetry of the field equations since it contains duality transformations. However, the subgroup $SO(8)$ does not contain duality transformations and is a symmetry of the Lagrangian.

To construct a gauged theory we first modify the derivatives on the fields so that they transform covariantly under the local $SO(8)$. The fields which transform under $SO(8) \subset E_{7(+7)}$ are the vector and scalar fields. The derivatives on the vector fields appear through their field strengths $F_{\mu\nu}{}^{IJ}$, which we modify to the $SO(8)$ Yang–Mills field strengths

$$F_{\mu\nu}{}^{IJ} = 2\partial_{[\mu}A_{\nu]}{}^{IJ} \quad \longrightarrow \quad F_{\mu\nu}{}^{IJ} = 2\partial_{[\mu}A_{\nu]}{}^{IJ} + 2gA_{[\mu}{}^{IK}A_{\nu]}{}^{KJ}. \tag{7.1}$$

Here, we have introduced a gauge coupling constant g as a new parameter. The derivatives on the scalar fields appear through $P_{\mu ijkl}$ and $Q_\mu{}^i{}_j$ in (4.66). We modify the derivatives there by including minimal couplings to the $SO(8)$ gauge fields as

$$\partial_\mu u^{IJ}{}_{ij} \quad \longrightarrow \quad \partial_\mu u^{IJ}{}_{ij} + 2gA_\mu{}^{K[I}u^{J]K}{}_{ij},$$
$$\partial_\mu v_{IJij} \quad \longrightarrow \quad \partial_\mu v_{IJij} + 2gA_\mu{}^K{}_{[I}v_{J]Kij}. \tag{7.2}$$

The new $P_{\mu ijkl}$ and $Q_\mu{}^i{}_j$ transform covariantly under the local $SO(8)$ and transform in the same way as before under the local $SU(8)$. The fermionic fields in the ungauged theory do not transform under $E_{7(+7)}$ and therefore under its subgroup $SO(8)$ neither. Hence, they have no direct minimal couplings with the $SO(8)$ gauge fields. However, the covariant derivatives on the fermionic fields contain the $SU(8)$ gauge field $Q_\mu{}^i{}_j$, which in turn depends on the $SO(8)$ gauge fields as in (7.2). Therefore, there are couplings of the fermionic fields to the vector fields.

The Lagrangian and the supertransformation of the gauged theory can be written as

$$\mathcal{L} = \mathcal{L}_0 + \mathcal{L}_g, \qquad \delta_Q = \delta_{Q0} + \delta_{Qg}. \tag{7.3}$$

$\mathcal{L}_0$ and δ_{Q0} are obtained from the Lagrangian (4.67) and the supertransformation (4.72) in the ungauged theory by replacing the covariant derivatives and the field strength with the above $SO(8)$ covariant ones. These replacements destroy local supersymmetry of the ungauged theory. To recover local supersymmetry we have to add new terms $\mathcal{L}_g$ and δ_{Qg} to the Lagrangian and the supertransformation as in (7.3). With an appropriate choice of these additional terms we can obtain a theory with local supersymmetry. The terms to be added to the Lagrangian and the supertransformation are

$$\mathcal{L}_g = 4g^2 e \left(\frac{3}{4} |A_1^{ij}|^2 - \frac{1}{24} |A_2^i{}_{jkl}|^2 \right) + ge \left(A_1^{ij} \bar{\psi}_{\mu i} \gamma^{\mu\nu} \psi_{\nu j} \right.$$

$$\left. + \frac{1}{3\sqrt{2}} A_2^i{}_{jkl} \bar{\psi}_{\mu i} \gamma^\mu \lambda^{jkl} + A_3^{ijk,lmn} \bar{\lambda}_{ijk} \lambda_{lmn} + \text{c.c.} \right),$$

$$\delta_{Qg} \psi_\mu{}^i = g A_1^{ij} \gamma_\mu \varepsilon_j, \qquad \delta_{Qg} \lambda^{ijk} = -\sqrt{2} g \left(A_2^l{}_{ijk} \right)^* \varepsilon^l, \tag{7.4}$$

where A_1, A_2, A_3 are functions of the scalar fields. $\mathcal{L}_g$ consists of the scalar field potential and the Yukawa couplings. To preserve local supersymmetry the functions of the scalar fields are chosen as

$$A_1^{ij} = -\frac{1}{42} T_k^{ikj}, \quad A_2^i{}_{jkl} = -\frac{1}{6} T^i{}_{[jkl]}, \quad A_3^{ijk,lmn} = \frac{1}{432} \eta \varepsilon^{ijkpqr[lm} T^{n]}{}_{pqr}. \tag{7.5}$$

Here, we have introduced the T tensor defined by

$$T_i^{jkl} = \left(u_{IJ}{}^{kl} + v^{IJkl} \right) \left(u^{JK}{}_{im} u_{KI}{}^{jm} - v_{JKim} v^{KIjm} \right), \quad T^i{}_{jkl} = \left(T_i^{jkl} \right)^*. \tag{7.6}$$

Choosing $\mathcal{L}_g$ and δ_{Qg} in this way the modified Lagrangian is invariant up to total divergences under the modified local supertransformation. To show the invariance we have to use various identities of the T tensor, which can be derived from the fact that the scalar fields u, v are matrix elements of $E_{7(+7)}$. See the original work [3] for details.

The local symmetries of the gauged theory are those of the general coordinate transformation, the local Lorentz transformation, the local $SO(8)$ transformation, the local $SU(8)$ transformation and the local supertransformation. The global $E_{7(+7)}$ symmetry of the ungauged theory is broken since only the subgroup $SO(8)$ is gauged. The commutator of two local supertransformations remains the same form as (4.73) for the ungauged theory though δ_g now represents the $SO(8)$ gauge transformation. The transformation parameters on the right-hand side of (4.73) are also modified. $F_{\mu\nu}^{IJ}$ and $Q_\mu{}^i{}_j$ of (4.74) become the $SO(8)$ covariant ones. Furthermore, the parameters of the local Lorentz and local $SU(8)$ transformations get additional terms

$$\Delta \lambda_{ab} = -\frac{1}{2} g A_1^{ij} \bar{\varepsilon}_{2i} \gamma_{ab} \varepsilon_{1j} + \text{c.c},$$

$$\Delta \Lambda^i{}_j = -\frac{1}{6} g \left(T^i{}_{jkl} \bar{\varepsilon}_2^k \varepsilon_1^l + T_j^{ikl} \bar{\varepsilon}_{2k} \varepsilon_{1l} \right). \tag{7.7}$$

Here, we have considered the gauging of the $SO(8)$ subgroup of $E_{7(+7)}$. It is possible to gauge other subgroups of $E_{7(+7)}$. In fact, theories were constructed in which $SO(p,q)$ $(p + q = 8)$ or non-semisimple groups obtained from them by Inönü–Wigner contractions are gauged [8, 9]. In general, a choice of a subgroup of the global symmetry G of the ungauged theories to be gauged is not unique. There is a systematic method to find which subgroup can be gauged for each supergravity [17].

A Solution of the Field Equations

Let us consider a solution of the field equations. We assume that the vector and spinor fields vanish and the scalar fields are constants independent of the spacetime coordinates. The constant value of the scalar fields is determined by a stationary point of the potential. For one of the stationary points the scalar field matrix V in (4.64) is a unit matrix, i.e.,

$$u^{IJ}{}_{ij} = 2\delta_i^{[I}\delta_j^{J]}, \qquad v_{IJij} = 0. \tag{7.8}$$

We will show that this is a stationary point of the potential when we study fluctuations around this solution below. The gravitational field equation then becomes

$$R_{\mu\nu} = -12g^2 g_{\mu\nu}. \tag{7.9}$$

A solution of this equation is four-dimensional anti de Sitter (AdS$_4$) spacetime with the inverse radius $m = 2g$ as we explained in Sect. 2.5. AdS$_4$ spacetime has the $SO(2, 3)$ symmetry as the isometry. The fields other than the gravitational field are also invariant under the $SO(2, 3)$ transformation. Therefore, the whole solution has the $SO(2, 3)$ symmetry. This solution is also invariant under the global diagonal $SO(8)$ subgroup of local $SO(8) \times SU(8)$ (simultaneous transformations of $SO(8)$ and $SO(8)$ subgroup of $SU(8)$).

Furthermore, this solution has global supersymmetry. To see supersymmetry of the solution, we need to show that the variation of the solution under the supertransformation in (7.3) vanishes. Since the fermionic fields vanish, supertransformations of the bosonic fields automatically vanish. Supertransformations of the fermionic fields are

$$\delta_Q \psi_\mu{}^i = D_\mu \varepsilon^i - g\gamma_\mu \varepsilon_i, \qquad \delta_Q \lambda^{ijk} = 0, \tag{7.10}$$

where we have used the fact that $A_1^{ij} = -\delta^{ij}$, $A_2{}^i{}_{jkl} = 0$ for this solution. λ^{ijk} is automatically invariant. The invariance of $\psi_\mu{}^i$ requires a Killing spinor equation $D_\mu \varepsilon^i = g\gamma_\mu \varepsilon_i$ on the transformation parameter ε^i. The charge conjugation of this equation is $D_\mu \varepsilon_i = g\gamma_\mu \varepsilon^i$. The integrability condition of these equations then becomes

$$[D_\mu, D_\nu]\varepsilon^i = D_\mu(g\gamma_\nu \varepsilon_i) - (\mu \leftrightarrow \nu) = -2g^2 \gamma_{\mu\nu}\varepsilon^i. \tag{7.11}$$

Using the general formula (1.30) we find

$$\left(\frac{1}{4}R_{\mu\nu}{}^{ab}\gamma_{ab} + 2g^2\gamma_{\mu\nu}\right)\varepsilon^i = 0. \tag{7.12}$$

This condition is satisfied by AdS$_4$ spacetime with the inverse radius $m = 2g$ due to (2.45). Thus, this solution has the global $SO(2, 3) \times SO(8)$ symmetry as well as the global $\mathcal{N} = 8$ supersymmetry. These global symmetry transformations have closed commutation relations and form the super AdS algebra $OSp(8|4)$ (See Sect. 3.5).

Next, we shall consider small fluctuations of the scalar fields around this solution as a background. To study the fluctuations it is convenient to fix a gauge of the local $SU(8)$. We use the gauge fixing condition

$$V(x) = \exp\begin{pmatrix} 0 & \frac{1}{\sqrt{2}}(\phi_{ijkl})^*(x) \\ \frac{1}{\sqrt{2}}\phi_{ijkl}(x) & 0 \end{pmatrix}, \quad \phi_{ijkl} = \frac{1}{4!}\eta\varepsilon_{ijklmnpq}(\phi_{mnpq})^*. \tag{7.13}$$

The exponent is a matrix of the $E_{7(+7)}$ Lie algebra (4.61) with vanishing Λ, which corresponds to $SU(8)$ gauge degrees of freedom. The scalar fields ϕ represent the fluctuations. In this gauge u, v in (4.64) are

$$u^{ij}{}_{kl} = 2\delta^{[i}_k \delta^{j]}_l + \frac{1}{8}\phi^{ijmn}\phi_{mnkl} + \mathcal{O}(\phi^4), \quad v_{ijkl} = \frac{1}{\sqrt{2}}\phi_{ijkl} + \mathcal{O}(\phi^3). \tag{7.14}$$

Then, we find the Lagrangian for the small fluctuations as

$$\mathscr{L} = -\frac{1}{48}e\partial_\mu\phi_{ijkl}\partial^\mu\phi^{ijkl} + \frac{1}{6}g^2 e\phi_{ijkl}\phi^{ijkl} + \mathcal{O}(\phi^3). \tag{7.15}$$

Since there is no linear term in the fluctuations, the solution (7.8) indeed corresponds to a stationary point of the potential. However, this stationary point is not a minimum but a maximum of the potential as can be seen from the coefficient of the ϕ^2 term. Accordingly, the mass square of the fluctuations has a negative value $m_\phi^2 = -2m^2$, where $m = 2g$ is the inverse radius of AdS spacetime. In Minkowski spacetime this implies that this solution is unstable. However, in AdS spacetime it is stable if the mass square is larger than a certain negative value. In D-dimensional AdS spacetime with the inverse radius m the condition of stability for a scalar field with the mass m_ϕ is [2, 11]

$$m_\phi^2 \geq -\frac{1}{4}(D-1)^2 m^2. \tag{7.16}$$

The mass in (7.15) satisfies this condition and therefore this solution is stable.

As we saw in Sect. 6.3, $D = 4$, $\mathcal{N} = 8$ Poincaré supergravity can be obtained from $D = 11$ supergravity by a dimensional reduction. One compactifies the $D = 11$ theory to $D = 4$ by a seven-dimensional torus and then keeps only massless modes discarding massive Kaluza–Klein modes. $D = 4$, $\mathcal{N} = 8$ gauged supergravity is also related to a compactification of $D = 11$ supergravity. In this case one compactifies the $D = 11$ theory to $D = 4$ by a seven-dimensional sphere S^7 and keeps only fields in the same supermultiplet as the four-dimensional gravitational field discarding other Kaluza–Klein modes [4, 5]. $D = 11$ supergravity has $AdS_4 \times S^7$ spacetime as a solution of the field equations. This solution corresponds to the AdS_4 solution of $D = 4$, $\mathcal{N} = 8$ gauged supergravity discussed above.

7.3 Gauged Supergravities in Higher Dimensions

One can construct a gauged supergravity in higher dimensions by gauging a subgroup of the global symmetry G of a Poincaré supergravity in Table 4.1. When antisymmetric tensor fields are contained in the theory, however, one has to be careful about how to treat them. Let us explain this point by considering two examples in higher dimensions.

7.3.1 $D = 7$, $\mathcal{N} = 4$ Gauged Supergravity

The bosonic fields of the ungauged theory are a gravitational field, 10 vector fields, 5 second rank antisymmetric tensor fields and 14 scalar fields as shown in Table 3.3. The Lagrangian has the global non-compact $SL(5, \mathbb{R})$ symmetry and the local $SO(5)$ symmetry as shown in Table 4.1. The scalar fields are described by the $SL(5, \mathbb{R})/SO(5)$ non-linear sigma model. The vector fields $A_\mu{}^{IJ}(= -A_\mu{}^{JI})$ and antisymmetric tensor fields $B_{\mu\nu}^I$ ($I, J = 1, \ldots, 5$) belong to the representations **10** and **5** of $SL(5, \mathbb{R})$, respectively.

We can consider a gauging of an $SO(5)$ subgroup of $SL(5, \mathbb{R})$ by using these ten vector fields. However, there is a difficulty to achieve this; the $SO(5)$ gauge symmetry of the vector fields is incompatible with the gauge symmetry of the antisymmetric tensor fields. Since the antisymmetric tensor fields transform under $SO(5) \in SL(5.\mathbb{R})$ in a nontrivial way, their $SO(5)$ covariant field strengths should be defined as

$$F_{\mu\nu\rho}^I = 3D_{[\mu}B_{\nu\rho]}^I \equiv 3\left(\partial_{[\mu}B_{\nu\rho]}^I + gA_{[\mu}^{IJ}B_{\nu\rho]}^J\right) \tag{7.17}$$

by using the $SO(5)$ covariant derivatives. However, these field strengths are not invariant under the gauge transformation of the antisymmetric tensors $\delta_g B_{\mu\nu}^I = 2D_{[\mu}\zeta_{\nu]}^I$ but transform as

$$\delta_g F_{\mu\nu\rho}^I = 3F_{[\mu\nu}^{IJ}\zeta_{\rho]}^J. \tag{7.18}$$

Therefore, the kinetic term of the antisymmetric tensor fields is not invariant under this transformation.

This difficulty was overcome as follows [13]. First, the second rank antisymmetric tensor fields $B_{\mu\nu}^I$ are replaced by dual third rank antisymmetric tensor fields $\tilde{B}_{\mu\nu\rho}^I$. As discussed in Sect. 1.4, these two kinds of fields with the Maxwell type Lagrangians are equivalent. Next, the massless Maxwell type Lagrangian of $\tilde{B}^I_{\mu\nu\rho}$ is replaced by a massive Chern–Simons type Lagrangian as discussed in Sect. 1.4. The third rank antisymmetric tensor fields with these two kinds of Lagrangians have the same physical degrees of freedom. Therefore, the physical degrees of freedom of the bosonic and fermionic fields remain the same. Since the massive Chern–Simons type Lagrangian does not have gauge invariance from the start, one need not worry about their gauge invariance when introducing the $SO(8)$ covariant derivatives.

In this way $D = 7$, $\mathcal{N} = 4$ gauged supergravity was constructed in [13]. The kinetic term of the third rank antisymmetric tensor fields has a form of massive Chern–Simons type using the $SO(5)$ covariant derivative

$$\mathscr{L} = \frac{1}{72} m \varepsilon^{\mu\nu\rho\sigma\tau\lambda\kappa} B^I_{\mu\nu\rho} D_\sigma B^I_{\tau\lambda\kappa} - \frac{1}{12} m^2 e V_i{}^I V_i{}^J B^I_{\mu\nu\rho} B^{J\mu\nu\rho}. \tag{7.19}$$

Here, $V_i{}^I$ is the inverse matrix of the scalar fields $V_I{}^i$ in $SL(5, \mathbb{R})$. The mass parameter m is related to the gauge coupling constant g as $m = \frac{1}{\sqrt{2}} g$.

The existence of the third rank antisymmetric tensor fields with the Chern–Simons type Lagrangian in this theory was suggested by a compactification from higher dimensions. $D = 11$, $\mathcal{N} = 1$ supergravity has a solution $AdS_7 \times S^4$. Compactifying the $D = 11$ theory to seven dimensions by S^4 one obtains $D = 7$, $\mathcal{N} = 4$ gauged supergravity [12]. The gauge group is $SO(5)$, which is the isometry of S^4. By studying small fluctuations around the background $AdS_7 \times S^4$, third rank antisymmetric tensor fields with massive Chern–Simons type field equations were found.

7.3.2 $D = 5$, $\mathcal{N} = 8$ Gauged Supergravity

The bosonic fields of the ungauged theory are a gravitational field, 27 vector fields and 42 scalar fields as shown in Table 3.3. The Lagrangian has the global non-compact $E_{6(+6)}$ symmetry and the local $USp(8)$ symmetry as shown in Table 4.1. The scalar fields are described by the $E_{6(+6)}/USp(8)$ non-linear sigma model. The vector fields belong to the representation **27** of $E_{6(+6)}$. Since there is no 27-dimensional semisimple Lie algebra, it is not clear which subgroup of $E_{6(+6)}$ should be gauged by using these vector fields.

A hint for constructing a gauged theory came from a compactification of a higher dimensional theory. $D = 10$, $\mathcal{N} = (2, 0)$ supergravity has a solution $AdS_5 \times S^5$. Compactifying this $D = 10$ theory to five dimensions by S^5 one expects to obtain $D = 5$, $\mathcal{N} = 8$ gauged supergravity. Studying small fluctuations around the background $AdS_5 \times S^5$ one found 15 massless vector fields and 12 second rank antisymmetric tensor fields. The second rank antisymmetric tensor fields satisfy massive Chern–Simons type field equations. Since S^5 has the isometry $SO(6)$, 15 vector fields are $SO(6)$ gauge fields. The second rank antisymmetric tensor fields have the same degrees of freedom as massless vector fields with the Maxwell type Lagrangian as discussed in Sect. 1.4.

These observations suggest that one should replace 12 vector fields out of 27 with the second rank antisymmetric tensor fields with a massive Chern–Simons type Lagrangian. The remaining 15 vector fields can be used to gauge a subgroup $SO(6)$ of $E_{6(+6)}$. In fact, such a gauged theory was constructed in [6, 14]. Since $SO(6)$ to be gauged is a subgroup of $SL(6, \mathbb{R})$ in $SL(6, \mathbb{R}) \times SL(2, \mathbb{R}) \in E_{6(+6)}$, the gauged theory still has a global $SL(2, \mathbb{R})$ symmetry.

7.4 $D = 10$, $\mathscr{N} = (1, 1)$ Massive Supergravity

There are two kinds of deformations of $D = 10$, $\mathscr{N} = (1, 1)$ Poincaré supergravity discussed in Sect. 5.3. One is the massive supergravity without minimal gauge couplings [16] and the other is the gauged supergravity with minimal gauge couplings [1, 7, 10]. In this section we shall discuss the massive supergravity.

It is possible to introduce a parameter m into $D = 10$, $\mathscr{N} = (1, 1)$ Poincaré supergravity and to construct a theory which has a scalar field potential and Yukawa couplings. This theory does not have minimal couplings to the vector field but has a similarity with gauged supergravities in other aspects. It is called $D = 10$, $\mathscr{N} = (1, 1)$ massive supergravity. The Poincaré supergravity in Sect. 5.3 will be called the massless theory below.

The field content of the massive theory is the same as that of the massless theory: a gravitational field $e_\mu{}^a(x)$, real antisymmetric tensor fields $B_{\mu\nu\rho}(x)$, $B_{\mu\nu}(x)$, a real vector field $B_\mu(x)$, a real scalar field $\phi(x)$, a Majorana Rarita–Schwinger field $\psi_\mu(x)$ and a Majorana spinor field $\lambda(x)$. The Lagrangian and the local supertransformation can be written as

$$\mathscr{L} = \mathscr{L}_0 + \mathscr{L}_m, \qquad \delta_Q = \delta_{Q0} + \delta_{Qm}. \tag{7.20}$$

$\mathscr{L}_0$ and δ_{Q0} are the Lagrangian and the supertransformation of the massless theory with the field strengths of the vector and antisymmetric tensor fields replaced by

$$F_{\mu\nu} = 2\partial_{[\mu} B_{\nu]} + m B_{\mu\nu}, \qquad F_{\mu\nu\rho} = 3\partial_{[\mu} B_{\nu\rho]},$$
$$F_{\mu\nu\rho\sigma} = 4\partial_{[\mu} B_{\nu\rho\sigma]} + 4B_{[\mu} F_{\nu\rho\sigma]} + 3m B_{[\mu\nu} B_{\rho\sigma]}. \tag{7.21}$$

$\mathscr{L}_m$ consists of a potential of the scalar fields, Yukawa couplings and a Chern–Simons term

$$\mathscr{L}_m = -\frac{1}{2} m^2 e\, e^{\frac{5}{2}\phi} + \frac{1}{8} m e\, e^{\frac{5}{4}\phi} \bar{\psi}_\mu \gamma^{\mu\nu} \psi_\nu - \frac{5}{8\sqrt{2}} m e\, e^{\frac{5}{4}\phi} \bar{\psi}_\mu \gamma^\mu \lambda$$
$$- \frac{21}{32} m e\, e^{\frac{5}{4}\phi} \bar{\lambda}\lambda - \frac{1}{288} \varepsilon^{\mu_1 \cdots \mu_{10}} \left(m \partial_{\mu_1} B_{\mu_2\mu_3\mu_4} B_{\mu_5\mu_6} B_{\mu_7\mu_8} B_{\mu_9\mu_{10}} \right.$$
$$\left. + \frac{9}{40} m^2 B_{\mu_1\mu_2} B_{\mu_3\mu_4} B_{\mu_5\mu_6} B_{\mu_7\mu_8} B_{\mu_9\mu_{10}} \right). \tag{7.22}$$

δ_{Qm} only contributes to the transformation of the fermionic fields

$$\delta_{Qm}\psi_\mu = \frac{1}{32}m\, e^{\frac{5}{4}\phi}\gamma_\mu\varepsilon, \qquad \delta_{Qm}\lambda = \frac{5}{8\sqrt{2}}m\, e^{\frac{5}{4}\phi}\varepsilon. \tag{7.23}$$

There is no minimal couplings to the vector field B_μ, and this theory is not a gauged supergravity.

The Lagrangian is invariant up to total divergences under the general coordinate transformation, the local Lorentz transformation, the gauge transformations of the vector and antisymmetric tensor fields and the local supertransformation as in the massless theory. The gauge transformations in (5.9) are replaced by

$$\delta_g B_\mu = \partial_\mu\zeta - m\zeta_\mu, \qquad \delta_g B_{\mu\nu} = 2\partial_{[\mu}\zeta_{\nu]},$$
$$\delta_g B_{\mu\nu\rho} = 3\partial_{[\mu}\zeta_{\nu\rho]} + 3B_{[\mu\nu}\partial_{\rho]}\zeta - 3m B_{[\mu\nu}\zeta_{\rho]}. \tag{7.24}$$

The field strengths (7.21) are invariant under these gauge transformations. The commutator of two local supertransformations has the same form as (5.11) of the massless theory but the transformation parameters on the right-hand side in (5.12) are modified; The field strengths of the vector and antisymmetric tensor fields are replaced by (7.21), and the parameter of the local Lorentz transformation has an additional term

$$\Delta\lambda_{ab} = -\frac{1}{64}m\, e^{\frac{5}{4}\phi}\bar\varepsilon_2\gamma_{ab}\varepsilon_1. \tag{7.25}$$

The global $\mathbb{R}^+$ symmetry of the massless theory is broken by the deformation.

When $m \neq 0$, we can use the gauge transformation with the parameter ζ_μ in (7.24) to set $B_\mu = 0$. In this gauge the kinetic term of the vector field becomes

$$-\frac{1}{4}e\, e^{\frac{3}{2}\phi}F_{\mu\nu}F^{\mu\nu} = -\frac{1}{4}m^2 e\, e^{\frac{3}{2}\phi}B_{\mu\nu}B^{\mu\nu}. \tag{7.26}$$

When ϕ has a constant background value, this term represents a mass term of the field $B_{\mu\nu}$. This is the Stückelberg mechanism similar to the Higgs mechanism. The antisymmetric tensor field $B_{\mu\nu}$ absorbs the vector field B_μ as longitudinal modes and becomes massive.

Because of the scalar field potential of the exponential form, this theory does not have 10-dimensional Minkowski spacetime as a solution of the field equations. Instead, this theory has an 8-brane solution, which corresponds to D8 branes in type IIA superstring theory. The parameter m represents the magnitude of the field strength $F_{\mu_1\cdots\mu_{10}} = me\varepsilon_{\mu_1\cdots\mu_{10}}$ of the ninth rank antisymmetric tensor field coupled to the D8-branes [15].

References

1. E. Bergshoeff, T. de Wit, U. Gran, R. Linares, D. Roest, (Non)Abelian gauged supergravities in nine-dimensions. JHEP **10**, 061 (2002). [hep-th/0209205]
2. P. Breitenlohner, D.Z. Freedman, Stability in gauged extended supergravity. Ann. Phys. **144**, 249 (1982)
3. B. de Wit, H. Nicolai, $\mathcal{N} = 8$ supergravity. Nucl. Phys. **B208**, 323 (1982)
4. B. de Wit, H. Nicolai, The consistency of the S^7 truncation in $D = 11$ supergravity. Nucl. Phys. **B281**, 211 (1987)
5. M.J. Duff, B.E.W. Nilsson, C.N. Pope, Kaluza–Klein supergravity. Phys. Rept. **130**, 1 (1986)
6. M. Günaydin, L.J. Romans, N.P. Warner, Gauged $\mathcal{N} = 8$ supergravity in five-dimensions. Phys. Lett. **B154**, 268 (1985)
7. P.S. Howe, N.D. Lambert, P.C. West, A new massive type IIA supergravity from compactification. Phys. Lett. B **416**, 303 (1998). [hep-th/9707139]
8. C.M. Hull, Noncompact gaugings of $\mathcal{N} = 8$ supergravity. Phys. Lett. **B142**, 39 (1984)
9. C.M. Hull, More gaugings of $\mathcal{N} = 8$ supergravity. Phys. Lett. **B148**, 297 (1984)
10. I.V. Lavrinenko, H. Lü, C.N. Pope, Fiber bundles and generalized dimensional reduction. Class. Quant. Grav. **15**, 2239 (1998). [hep-th/9710243]
11. L. Mezincescu, P.K. Townsend, Stability at a local maximum in higher dimensional anti-de Sitter space and applications to supergravity. Ann. Phys. **160**, 406 (1985)
12. H. Nastase, D. Vaman, P. van Nieuwenhuizen, Consistent nonlinear KK reduction of 11d supergravity on AdS$_7 \times S^4$ and selfduality in odd dimensions. Phys. Lett. **B469**, 96 (1999). [hep-th/9905075]
13. M. Pernici, K. Pilch, P. van Nieuwenhuizen, Gauged maximally extended supergravity in seven-dimensions. Phys. Lett. **B143**, 103 (1984)
14. M. Pernici, K. Pilch, P. van Nieuwenhuizen, Gauged $\mathcal{N} = 8$ $D = 5$ supergravity. Nucl. Phys. **B259**, 460 (1985)
15. J. Polchinski, Dirichlet branes and Ramond-Ramond charges. Phys. Rev. Lett. **75**, 4724 (1995). [hep-th/9510017]
16. L.J. Romans, Massive $\mathcal{N} = 2a$ supergravity in ten-dimensions. Phys. Lett. **B169**, 374 (1986)
17. H. Samtleben, Lectures on gauged supergravity and flux compactifications. Class. Quant. Grav. **25**, 214002 (2008). arXiv:0808.4076 [hep-th]

Appendix A
Notation and Conventions

We use natural units, in which the speed of light is $c = 1$ and Planck's constant is $\hbar = 1$. We also take Newton's gravitational constant as $16\pi G = 1$. In this system of units the Einstein term for the gravitational field in the Lagrangian has a coefficient 1. One can recover c, $\hbar$ and G in other system of units by dimensional analysis.

We use $\mu, \nu, \rho, \ldots = 0, 1, 2, \ldots, D - 1$ for world indices of D-dimensional spacetime and $a, b, c, \ldots = 0, 1, 2, \ldots, D - 1$ for local Lorentz indices. The flat metric in a local Lorentz frame is

$$\eta_{ab} = \text{diag}(-1, +1, \ldots, +1) = \eta^{ab}. \tag{A.1}$$

The Levi-Civita symbols $\varepsilon^{\mu_1\mu_2\ldots\mu_D}$ and $\varepsilon_{\mu_1\mu_2\ldots\mu_D}$ represent totally antisymmetric tensor densities and have components

$$\varepsilon^{012\ldots D-1} = +1, \quad \varepsilon_{012\ldots D-1} = -1. \tag{A.2}$$

We also use the Levi-Civita symbols $\varepsilon^{a_1a_2\ldots a_D}$ and $\varepsilon_{a_1a_2\ldots a_D}$ with local Lorentz indices, which are totally antisymmetric and have the same components as (A.2).

Symmetrization and antisymmetrization of n indices with unit strength are denoted as

$$T_{(a_1\ldots a_n)} = \frac{1}{n!} \sum_P T_{P(a_1\ldots a_n)},$$

$$T_{[a_1\ldots a_n]} = \frac{1}{n!} \sum_P (-1)^P T_{P(a_1\ldots a_n)}, \tag{A.3}$$

where $\sum_P$ is a sum over permutations $P(a_1 \ldots a_n)$ of the indices $a_1 a_2 \ldots a_n$. The sign factor $(-1)^P$ is $+1$ for even permutations and -1 for odd permutations. For instance,

$$T_{(ab)} = \frac{1}{2}(T_{ab} + T_{ba}), \quad T_{[ab]} = \frac{1}{2}(T_{ab} - T_{ba}). \tag{A.4}$$

Y. Tanii, *Introduction to Supergravity*, SpringerBriefs in Mathematical Physics, DOI: 10.1007/978-4-431-54828-7, © The Author(s) 2014

For conventions of geometric quantities such as curvatures see Sect. 1.2.

Groups

Here, we summarize definitions of groups appearing in the text. In the following m and n denote natural numbers, and we define

$$\eta = \begin{pmatrix} -\mathbf{1}_{m \times m} & 0 \\ 0 & \mathbf{1}_{n \times n} \end{pmatrix} = \eta^T, \quad \Omega = \begin{pmatrix} 0 & \mathbf{1}_{n \times n} \\ -\mathbf{1}_{n \times n} & 0 \end{pmatrix} = -\Omega^T, \tag{A.5}$$

where $\mathbf{1}_{n \times n}$ is the $n \times n$ unit matrix.

- $GL(n, \mathbb{R})$: general linear group
  The group of $n \times n$ real matrices with a non-vanishing determinant. A non-compact group of dimension n^2. $GL(1, \mathbb{R})^+ = \mathbb{R}^+$ is a group of positive real numbers.
- $SL(n, \mathbb{R})$: special linear group
  The group of $n \times n$ real matrices with a unit determinant. A non-compact group of dimension $n^2 - 1$.
- $SO(m, n)$: special orthogonal group
  The group of $(m + n) \times (m + n)$ real matrices O satisfying

$$O^T \eta O = \eta, \quad \det O = 1. \tag{A.6}$$

$SO(0, n) = SO(n, 0) = SO(n)$ is the n-dimensional rotation group, and $SO(1, D - 1)$ is the D-dimensional Lorentz group. $SO(n)$ is compact and $SO(m, n)$ $(m, n \neq 0)$ is non-compact. The dimension is $\frac{1}{2}(m + n)(m + n - 1)$.

- $SU(m, n)$: special unitary group
  The group of $n \times n$ complex matrices U satisfying

$$U^\dagger \eta U = \eta, \quad \det U = 1. \tag{A.7}$$

$SU(0, n) = SU(n, 0) = SU(n)$ is a compact special unitary group. $SU(m, n)$ $(m, n \neq 0)$ is non-compact. Dimension is $(m + n)^2 - 1$. When the condition $\det U = 1$ is not imposed, the group is called $U(m, n)$.

- $Sp(2n, \mathbb{R})$: real symplectic group
  The group of $2n \times 2n$ real matrices M satisfying

$$M^T \Omega M = \Omega. \tag{A.8}$$

A non-compact group of dimension $n(2n + 1)$.

- $USp(2n)$: unitary symplectic group
  The group of $2n \times 2n$ complex matrices U satisfying

$$U^T \Omega U = \Omega, \quad U^\dagger U = 1. \tag{A.9}$$

A compact group of dimension $n(2n + 1)$.

Appendix B
Formulae of Gamma Matrices

Antisymmetrized Products

The antisymmetrized products of n gamma matrices of $SO(t,s)$ $(t+s=D)$ are defined as

$$\gamma^{a_1 a_2 \cdots a_n} = \gamma^{[a_1} \gamma^{a_2} \cdots \gamma^{a_n]} \quad (n = 1, 2, \ldots, D). \tag{B.1}$$

When all the indices $a_1, \ldots, a_n$ are different, $\gamma^{a_1 \cdots a_n} = \gamma^{a_1} \gamma^{a_2} \cdots \gamma^{a_n}$. They are traceless $\mathrm{tr}\, \gamma^{a_1 a_2 \cdots a_n} = 0$.

Duality Relations

When D is even, the antisymmetrized products satisfy duality relations

$$\gamma^{a_1 \cdots a_n} = (-1)^{\frac{1}{4}(s-t)+\frac{1}{2}n(n-1)} \frac{1}{(D-n)!} \varepsilon^{a_1 \cdots a_D} \gamma_{a_{n+1} \cdots a_D} \bar{\gamma}, \tag{B.2}$$

where $\varepsilon^{a_1 \cdots a_D}$ is the totally antisymmetric Levi-Civita symbol and $\bar{\gamma}$ is the chirality matrix (3.11). When D is odd, they satisfy

$$\gamma^{a_1 \cdots a_n} = (-1)^{\frac{1}{4}(s+3t-1)+\frac{1}{2}n(n-1)} \frac{1}{(D-n)!} \varepsilon^{a_1 \cdots a_D} \gamma_{a_{n+1} \cdots a_D}. \tag{B.3}$$

Products

A product of two antisymmetrized products can be expressed as a linear combination of antisymmetrized products as

$$\gamma^{a_1...a_m} \gamma_{b_1...b_n} = \gamma^{a_1...a_m}{}_{b_1...b_n} + mn\, \delta^{[a_m}_{[b_1} \gamma^{a_1...a_{m-1}]}{}_{b_2...b_n]}$$

$$+ 2\,_mC_2\,_nC_2\, \delta^{[a_m}_{[b_1} \delta^{a_{m-1}}_{b_2} \gamma^{a_1...a_{m-2}]}{}_{b_3...b_n]}$$

$$+ 3!\,_mC_3\,_nC_3\, \delta^{[a_m}_{[b_1} \delta^{a_{m-1}}_{b_2} \delta^{a_{m-2}}_{b_3} \gamma^{a_1...a_{m-3}]}{}_{b_4...b_n]} + \cdots . \qquad (B.4)$$

The last term contains an antisymmetrized product of $|m - n|$ gamma matrices. For instance,

$$\gamma^{abc} \gamma_{de} = \gamma^{abc}{}_{de} + 6\gamma^{[ab}{}_{[e} \delta^{c]}_{d]} + 6\gamma^{[a} \delta^c_{[d} \delta^{b]}_{e]}. \qquad (B.5)$$

The first term of (B.4) corresponds to the case in which $a_1, \ldots, a_m$ and $b_1, \ldots, b_n$ do not contain the same number, the second term corresponds to the case in which only one of $a_1, \ldots, a_m$ is the same as one of $b_1, \ldots, b_n$, the third term corresponds to the case in which only two of $a_1, \ldots, a_m$ are the same as two of $b_1, \ldots, b_n$, and so on. The coefficients can be found by counting combinatorial numbers. From (B.4) we can obtain formulae such as

$$\gamma_b \gamma^{ba_1a_2...a_n} = (D - n)\gamma^{a_1a_2...a_n} = \gamma^{a_1a_2...a_n b} \gamma_b,$$

$$\gamma^b \gamma^{a_1a_2...a_n} \gamma_b = (-1)^n (D - 2n)\gamma^{a_1a_2...a_n}. \qquad (B.6)$$

Completeness

When D is even, a set of matrices

$$\mathbf{1}, \gamma^{a_1}, \gamma^{a_1a_2}, \ldots, \gamma^{a_1a_2...a_D}, \qquad (B.7)$$

where $a_1 < a_2 < \cdots$ for each matrix, are linearly independent and form a complete set of $2^{D/2} \times 2^{D/2}$ complex matrices. Namely, an arbitrary $2^{D/2} \times 2^{D/2}$ complex matrix can be expressed uniquely as a linear combination of these matrices. To show this, notice that the number of the matrices in (B.7) is

$$_DC_0 + {}_DC_1 + {}_DC_2 + \cdots + {}_DC_D = (1 + 1)^D = 2^D, \qquad (B.8)$$

which is equal to the number of components of a $2^{D/2} \times 2^{D/2}$ matrix. Furthermore, the matrices in (B.7) are orthogonal with respective to the trace

$$\mathrm{tr}\left(\gamma^{a_1...a_m} \gamma_{b_n...b_1}\right) = \begin{cases} 2^{D/2} \delta^{a_1}_{b_1} \ldots \delta^{a_m}_{b_m} & (m = n) \\ 0 & (m \neq n) \end{cases}, \qquad (B.9)$$

where $a_1 < a_2 < \cdots < a_m$, $b_1 < b_2 < \cdots < b_n$. Therefore, the matrices in (B.7) are a set of 2^D linearly independent matrices and hence form a complete set of $2^{D/2} \times 2^{D/2}$ matrices.

When D is odd, a set of matrices

$$\mathbf{1}, \quad \gamma^a, \quad \gamma^{a_1 a_2}, \ldots, \gamma^{a_1 a_2 \ldots a_{\frac{1}{2}(D-1)}} \tag{B.10}$$

are linearly independent and form a complete set of $2^{(D-1)/2} \times 2^{(D-1)/2}$ matrices. This can be shown by the above even dimensional result and the fact that odd D-dimensional gamma matrices can be constructed using even $(D-1)$-dimensional ones (See Sect. 3.1). Note that (B.10) does not contain $\gamma^{a_1 \ldots a_n}$ for $n > \frac{1}{2}(D-1)$, which are related to those for $n \leq \frac{1}{2}(D-1)$ as in (B.3) and are not independent.

Bilinear Forms

From two spinors ψ, λ we can construct a bilinear form

$$\bar{\psi} \gamma^{a_1 \ldots a_n} \lambda, \tag{B.11}$$

which transforms as an n-th rank tensor under $SO(t, s)$ transformations in (3.6). Here, $\bar{\psi}$ is the Dirac conjugate of ψ defined in (3.9).

Using the definition of the charge conjugation (3.22) and the properties of the gamma matrices (3.8), (3.23) we find

$$(\bar{\psi} \gamma^{a_1 \ldots a_n} \lambda)^\dagger = (-1)^{tn + \frac{1}{2}n(n-1)} \bar{\lambda} \gamma^{a_1 \ldots a_n} \psi,$$

$$\bar{\psi} \gamma^{a_1 \ldots a_n} \lambda = -(\pm 1)^{t+n} (-1)^{\frac{1}{2}t(t+1) + tn + \frac{1}{2}n(n-1)} \varepsilon_\pm \bar{\lambda}^c \gamma^{a_1 \ldots a_n} \psi^c. \tag{B.12}$$

Here, the sign factor ± 1 on the right-hand side of the second equation is $+1$ when the charge conjugation matrix C_+ is used and -1 when C_- is used. Components of the spinors are treated as anticommuting Grassmann numbers. For instance, for Majorana spinors $\psi = \psi^c$, $\lambda = \lambda^c$ of the four-dimensional Lorentz group $SO(1, 3)$ we find

$$(\bar{\psi} \lambda)^\dagger = \bar{\lambda} \psi = \bar{\psi} \lambda,$$

$$(\bar{\psi} \gamma^a \lambda)^\dagger = -\bar{\lambda} \gamma^a \psi = \bar{\psi} \gamma^a \lambda,$$

$$(\bar{\psi} \gamma^{ab} \lambda)^\dagger = -\bar{\lambda} \gamma^{ab} \psi = \bar{\psi} \gamma^{ab} \lambda,$$

$$(\bar{\psi} \gamma^{abc} \lambda)^\dagger = \bar{\lambda} \gamma^{abc} \psi = \bar{\psi} \gamma^{abc} \lambda,$$

$$(\bar{\psi} \gamma^{abcd} \lambda)^\dagger = \bar{\lambda} \gamma^{abcd} \psi = \bar{\psi} \gamma^{abcd} \lambda. \tag{B.13}$$

For Weyl spinors $\psi_\pm$, $\lambda_\pm$ of $SO(t, s)$ $(t + s = \text{even})$ we find

$$\bar{\psi}_\pm \gamma^{a_1 \ldots a_n} \lambda_\pm = 0 \quad (n + t = \text{odd}),$$

$$\bar{\psi}_\pm \gamma^{a_1 \ldots a_n} \lambda_\mp = 0 \quad (n + t = \text{even}), \tag{B.14}$$

which can be shown by (3.13), (3.14) and $\bar{\gamma} \gamma^{a_1 \ldots a_n} = (-1)^n \gamma^{a_1 \ldots a_n} \bar{\gamma}$.

Fierz Identities

A product of two bilinears $\bar{\psi}_1\psi_2$ and $\bar{\psi}_3\psi_4$ can be rearranged as

$$
\bar{\psi}_1\psi_2\bar{\psi}_3\psi_4 = -2^{-\left[\frac{D}{2}\right]}\Bigg[\bar{\psi}_1\psi_4\bar{\psi}_3\psi_2 + \bar{\psi}_1\gamma_a\psi_4\bar{\psi}_3\gamma^a\psi_2 + \frac{1}{2}\bar{\psi}_1\gamma_{ab}\psi_4\bar{\psi}_3\gamma^{ba}\psi_2
$$

$$
+ \frac{1}{3!}\bar{\psi}_1\gamma_{abc}\psi_4\bar{\psi}_3\gamma^{cba}\psi_2 + \cdots + \frac{1}{N!}\bar{\psi}_1\gamma_{a_1\ldots a_N}\psi_4\bar{\psi}_3\gamma^{a_N\ldots a_1}\psi_2 \Bigg],
$$

(B.15)

where $N = D$ when D is even and $N = \frac{1}{2}(D - 1)$ when D is odd. An identity like (B.15), which exchanges the combinations of the spinors, is called the Fierz identity.

The Fierz identity (B.15) can be shown as follows. First, the left-hand side can be written as

$$
\bar{\psi}_1\psi_2\bar{\psi}_3\psi_4 = \bar{\psi}_1^\alpha\psi_{2\beta}\bar{\psi}_3^\gamma\psi_{4\delta}\delta_\alpha^\beta\delta_\gamma^\delta,
$$

(B.16)

where $\alpha, \beta, \gamma, \delta = 1, 2, \ldots, 2^{[D/2]}$ are spinor indices. Regard $\delta_\alpha^\beta\delta_\gamma^\delta$ on the right-hand side of (B.16) as the $(\gamma\beta)$ component of a $2^{[D/2]} \times 2^{[D/2]}$ matrix for fixed α, δ. This matrix can be written as a linear combination of (B.7) or (B.10)

$$
\delta_\alpha^\beta\delta_\gamma^\delta = C_{0\alpha}{}^\delta\delta_\gamma^\beta + C_1{}_\alpha^{a\,\delta}(\gamma_a)_\gamma{}^\beta + C_2^{ab}{}_\alpha^{\,\delta}(\gamma_{ab})_\gamma{}^\beta + \cdots .
$$

(B.17)

The coefficients $C_n^{a_1\ldots a_n}{}_\alpha{}^\delta$ can be found by multiplying $(\gamma_{a_1\ldots a_n})_\beta{}^\gamma$ on the both-hand sides and using (B.9). In this way we find

$$
\delta_\alpha^\beta\delta_\gamma^\delta = 2^{-\left[\frac{D}{2}\right]}\Bigg[\delta_\alpha^\delta\delta_\gamma^\beta + (\gamma_a)_\alpha{}^\delta(\gamma^a)_\gamma{}^\beta + \frac{1}{2}(\gamma_{ab})_\alpha{}^\delta(\gamma^{ba})_\gamma{}^\beta
$$

$$
+ \frac{1}{3!}(\gamma_{abc})_\alpha{}^\delta(\gamma^{cba})_\gamma{}^\beta + \cdots + \frac{1}{N!}(\gamma_{a_1\ldots a_N})_\alpha{}^\delta(\gamma^{a_N\ldots a_1})_\gamma{}^\beta \Bigg].
$$

(B.18)

Substituting this into (B.16) and exchanging positions of the spinors we obtain (B.15). The minus sign on the right-hand side of (B.15) appears because of the Grassmann nature of the spinors.

Similarly, we can consider Fierz identities for other types of bilinears such as $\bar{\psi}_1\gamma^a\psi_2\bar{\psi}_3\gamma^{bc}\psi_4$. The identity in this case can be found by making replacements $\psi_2 \rightarrow \gamma^a\psi_2$, $\bar{\psi}_3 \rightarrow \bar{\psi}_3\gamma^{bc}$ in (B.15).

Index

A
Action, 2, 91
AdS spacetime, 27, 33, 111, 114
AdS supergravity, 27, 33, 111
   $D = 4$, $\mathcal{N} = 1$, 26
   $D = 4$, $\mathcal{N} = 2$, 32
Anomaly cancellation, 84
Anti de Sitter spacetime, *see* AdS spacetime
Anti de Sitter supergravity, *see* AdS supergravity
Antisymmetric tensor field, 9, 46, 58, 71, 96, 116
Antisymmetrization, 121
Automorphism, 43, 51
Auxiliary field, 20, 23

B
Bianchi identity, 9, 11, 32, 57, 58
Bilinear form, 125

C
Cartan subalgebra, 96, 105
Central charge, 28, 31, 43, 51, 73
Charge conjugation, 17, 40, 44, 65
Charge conjugation matrix, 17, 41
Chern–Simons term, 72, 84, 98, 104
Chern–Simons type, 13, 117
Chiral multiplet, 19, 20
Chirality, 39, 46, 73, 76, 83
Chirality matrix, 18, 39
Christoffel symbol, 3
Commutator algebra, 20, 23, 25, 27, 30, 33, 72, 75, 78, 85, 113, 119
Compactification, 89, 115, 117
Compensating transformation, 54, 81

Coset space, 53
Cosmological constant, 3, 27, 33
Cosmological term, 3, 26, 33, 111
Covariant derivative, 3, 6, 7, 9, 22, 33, 54, 112, 116
CPT theorem, 19, 29

D
D-brane, 1
D8 brane, 119
De Sitter spacetime, 27
Dilaton, 75, 82, 85
Dimensional reduction, 64, 90, 115
Dirac conjugate, 39
Dirac spinor, 39
Dual field, 11, 101–103, 106, 108, 116
Duality symmetry, 10, 32, 57
Duality transformation, 57, 100, 102, 103, 108

E
$E_{7(+7)}$, 64, 65, 112
Einstein equation, 3
Einstein frame, 76, 82, 86
Einstein term, 3, 76
Electric–magnetic duality, 57
Embedding tensor, 111
Energy–momentum tensor, 4, 11, 60
Extended super Poincaré algebra, 28
Extended supergravity, 29, 33
Extended supersymmetry, 28

F
Field strength, 6, 8, 9